ADVANCED
COMMON CORE
MATH
EXPLORATIONS

ADVANCED COMMON CORE MATH EXPLORATIONS

Measurement & Polygons

JERRY BURKHART

PRUFROCK PRESS INC.

WACO, TEXAS

Prufrock Press Inc.
P.O. Box 8813
Waco, TX 76714-8813
Phone: (800) 998-2208
Fax: (800) 240-0333
http://www.prufrock.com

Table of Contents

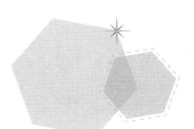

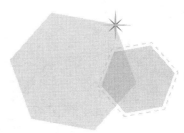

A Note to Students

Welcome, math explorers! You are about to embark on an adventure in learning. As you navigate the mathematical terrain in these activities, you will discover that "doing the math" means much more than calculating quickly and accurately. It means using your creativity and insight to question, investigate, describe, analyze, predict, and prove. It means venturing into unfamiliar territory, taking risks, and finding a way forward even when you're not sure which direction to go. And it means discovering things that will expand your mathematical imagination in entirely new directions.

Of course, the job of an explorer involves hard work. There may be times when it will take a real effort on your part to keep pushing forward. You may spend days or more pondering a single question or problem. Sometimes, you might even get completely lost. The process can be demanding—but it can also be very rewarding. There's nothing quite like the experience of making a breakthrough after a long stretch of hard work, and seeing a world of new ideas and understandings open up before your eyes!

These explorations are challenging, so you might want to team up with a partner or two on your travels—to discuss plans and strategies and to share the rewards of your hard work. Even if you don't reach your final destination every time, I believe you will find that the journey was worth taking. So, gear up for some adventure and hard work . . . and start exploring!

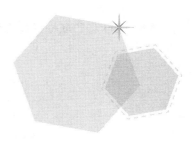

Introduction

This introduction contains general information about how the books and activities in the Advanced Common Core Math Explorations series are structured and how to use them. For additional information and support, please see the free e-booklet *Advanced Common Core Math Explorations: A Teacher's Guide* that accompanies this series at http://www.prufrock.com/Assets/ClientPages/AdvancedCCMath.aspx.

AUDIENCE

Advanced Common Core Math Explorations: Measurement and Polygons is designed to support students, teachers, and other learners as they work to deepen their understanding of middle school math concepts. The activities have been written primarily with upper elementary and middle school students and teachers in mind. However, older students or those who have already studied more advanced content can also enjoy and benefit from them. The explorations can be used in classrooms, as professional development activities for mathematics teachers, in college math content and methods courses, and by anyone who would like to extend his or her understanding of middle school mathematics concepts by solving challenging problems.

These explorations are designed to stretch students beyond their initial level of comfort. They are built around the belief that most of us underestimate the mathematics we are capable of learning. However, while the activities are challenging, they are also meant to be accessible. Although they are targeted to the special needs of gifted and talented students, I hope that teachers will make them available to any student who would like to pursue the challenge. Many students can make progress and learn something meaningful, even if they work just on the first question or two of an activity.

PURPOSE

The investigations in this series were developed through years of work with talented middle school math students. They are designed to:
   » engage students in the excitement of mathematical discovery;

» deepen students' understanding of a wide range of middle school math concepts;

» encourage the use of multiple strategies for solving problems;

» help students become flexible, creative, yet disciplined mathematical thinkers;

» improve mathematical communication skills;

» highlight connections between diverse mathematical concepts;

» develop perseverance, patience, and stamina in solving mathematical problems;

» provide levels of depth and challenge to meet a variety of needs and interests;

» enable students to work both collaboratively and independently; and

» offer opportunities for further exploration.

STRUCTURE OF THE BOOKS

Each book in the Advanced Common Core Math Explorations series contains ready-to-use explorations focused on one mathematical content area. The content and structure is built around the Common Core State Standards for Mathematics (National Governors Association Center for Best Practices & Council of Chief State School Officers [NGA & CCSSO], 2010), both the Content Standards and the Standards for Mathematical Practice. Because the emphasis is on challenge and depth, there is a stronger focus on concepts than on procedural skills. However, most activities provide plenty of opportunities to practice computational skills as well.

Each exploration is matched with one or more Common Core benchmarks or clusters, which come with grade-level designations. This grade-level information should be taken as a rough guide. When selecting activities, use your own knowledge of your students' backgrounds and abilities. Information about the prior knowledge needed for each exploration is also included as a guide.

FEATURES OF THE EXPLORATIONS

Each activity is divided into three stages. Stage 1 (and sometimes part of Stage 2) may be challenging enough to meet the needs of many students. The second and third stages are usually appropriate for older students, or for those who finish early, need more challenge, or are highly motivated and curious to learn more. They may also be useful for teachers or other adults who have more mathematical experience and want to extend their own knowledge further. I have separated the explorations into stages in order to provide a tool for setting goals, to help measure and celebrate students' progress, and to create additional options for those who need them.

Each exploration also contains features carefully designed to support teachers in the implementation process: an introduction, the student handout, a set of questions and notes to guide conversation, detailed solutions, and suggestions for a closing discussion.

IMPLEMENTING THE EXPLORATIONS

Implementing each exploration involves five steps on the part of the teacher: prepare, introduce, follow up, summarize, and assess.

Prepare

The best way to prepare to teach an activity is to try it yourself. Although this involves an initial time investment on your part, it pays great dividends later. Doing the activity, ideally with a partner or two, will help you become familiar with the mathematics, anticipate potential trouble spots for students, and plan ways to prepare students for success. After you've used the activity once or twice with students, very little preparation will be needed.

Introduce

The Introduction section at the beginning of each exploration provides support to help you get your students started: materials and prior knowledge needed, learning goals, motivational background, and suggestions for launching the activity.

Read the Motivation and Purpose selection to students, and then follow the suggestions for leading a discussion to help them understand the problem. Often, one of the suggestions involves looking through the entire activity with them (or as much of it as they will be doing) to help them see the big picture before they begin. Let students know what time frame you have in mind for the exploration. An activity may take anywhere from a few days to 2 or 3 weeks depending on how challenging it is for students, how much of it they will complete, and how much class time will be devoted to it.

The explorations are designed to allow students to spend much of their time working without direct assistance. However, it's usually best if you stay with them for a few minutes just after introducing an activity to ensure that they get started successfully. This way, you can catch potential trouble spots early and prevent unnecessary discouragement.

This is also a good time to remind students about the importance of giving clear, thorough written explanations of their thinking. Specific motivation techniques and suggestions for developing mathematical communication skills are included in the *Advanced Common Core Math Explorations: A Teacher's Guide* e-booklet.

Follow Up

The level of challenge in these explorations makes it impractical for most students to complete them entirely on their own as seatwork or homework. Students' most meaningful (and enjoyable!) experiences are often the opportunities you give them to have mathematical conversations with you and with each other while the activity is in progress. If you are implementing an activity with a small group of students in a mainstream classroom, it may be sufficient to plan to meet with them a couple of times per week, for 15 or 20 minutes each time. If circumstances allow you to spend more time than this, then the conversations and learning can be still better.

The Teacher's Guide for each exploration reprints each problem and contains two main elements: (a) Questions and Conversations and (b) Solutions. The Questions and Conversations feature is designed to help you facilitate these conversations with and among students. For the most part, it lists questions that students may ask or that you may pose to them. Ideas for responding to the questions are included. It isn't necessary to ask or answer all of the questions. Instead, let students' ideas and your experience and professional judgment determine the flow of the conversation. The Solutions feature offers ideas for follow-up discussions with students as they work. Although the answers in the Questions and Conversations sections are often intentionally incomplete or suggestive of ideas to consider, you will find detailed answers, often with samples of multiple approaches that students typically pursue, in the Solutions section.

Summarize

After students have finished an exploration, plan a brief discussion (20 minutes is usually enough) to give them a chance to share and critique one another's ideas and strategies. This is also a good time to answer any remaining questions they have. The Wrap Up section at the end of each exploration offers ideas for this discussion, along with suggestions for further exploration.

Assess

One of the most valuable things you can do for your students is to comment on their work. You do not have to write a lot, but your comments should show that you have read and thought about what they have written. Whether you give praise or offer suggestions for growth, make your comments specific and sincere. Ideally, some of your comments will relate to the detail of the mathematical content. Some specific suggestions are included in the free e-booklet accompanying this series.

If you would like to give students a numerical score, consider using a rubric such as the one in *Extending the Challenge in Mathematics: Developing Mathematical Promise in K–8 Students* (Sheffield, 2003). Whatever system you use, the emphasis should be on process goals such as problem solving, reasoning, communication, and making connections—not just correct answers. You may also build in general criteria such as effort, perseverance, correct spelling and grammar, organization, legibility, etc. However, remember that the central goal is to develop students' mathematical capacity. Any scoring system should reflect this.

GETTING STARTED

Below are some tips for getting started. First, a few DON'Ts to help you avoid some common pitfalls:

» *Don't feel that you have to finish the activities.* Students will learn more from thinking deeply about one or two questions than from rushing to finish an activity. Each exploration is designed to contain problems that will challenge virtually any student. Most students will not be able to answer every question.

» *Don't feel that you have to explain everything to students.* Your most important job is to help them learn to develop and test their own ideas. They will learn more if they do most of the thinking.

» *Don't be afraid to allow students to struggle.* Talented students need to know that meaningful learning takes time and hard work. Many of them need to experience some frustration—and learn to manage it.

» *Don't feel that you have to know all of the answers.* In order to challenge our students mathematically, we have to do the same for ourselves. You'll never know all of the answers, but if you're like me, you'll learn more about the math every time you teach an exploration! Do what you can during the time you've allotted to prepare, and then allow yourself to learn from the mathematical conversations—right along with your students.

And now some important DOs:

» *Take your time.* Allow the students plenty of time to think about the problems. Take the time to explore the ideas in depth rather than rushing to get to the next question.

» *Play with the mathematics!* To many people's surprise, math is very much about creative play. Of course, there are learning goals, and it takes effort, but be sure to enjoy playing with the patterns, numbers, shapes, and ideas!

» *Listen closely to students' ideas and expect them to listen closely to each other.* Meaningful mathematical conversation may be the single most important key to students' learning. It is also your key to assessing their learning.

» *Help students feel comfortable taking risks.* When you place less emphasis on the answers and show more interest in the quality of students' engagement, ideas, creativity, and questions, they will feel freer to make mistakes and grow from them.

» *Believe that the students—and you—can do it!* Middle school students have great success with these activities, but it may take some time to adjust to the level of challenge.

» *Use the explorations flexibly.* You don't always have to use them exactly "as is." Feel free to insert, delete, or modify questions to meet your students' needs. Adjust due dates or completion goals as necessary based on your observations of students.

Many teachers find it helpful to make a solid, but realistic commitment at the beginning of the school year to use the explorations. Put together a general plan for selecting students, forming groups, creating time for students to work (including time for you to meet with them), assessing the activities, and communicating with parents. Stick with your basic plan, making adjustments as needed as the school year progresses.

THE E-BOOKLET

The Advanced Common Core Math Explorations series comes with a free e-booklet (http://www.prufrock.com/Assets/ClientPages/AdvancedCCMath.aspx) that contains detailed suggestions and tools for bringing the activities to life in your classroom. It addresses topics such as motivation, questioning techniques, mathematical communication, assessment, parent communication, implementing the explorations in different settings, and identification.

Connections to the Common Core State Standards

COMMON CORE STATE STANDARDS FOR MATHEMATICAL CONTENT

Table 1 outlines connections between the activities in *Advanced Common Core Math Explorations: Measurement and Polygons* and the Common Core State Standards for Mathematics (NGA & CCSSO, 2010). The Standard column lists the CCSS Mathematical Content standards that apply to the activity. The Connections column shows other standards that are also addressed in the exploration. Extending the Core Learning shows how the activity extends student learning relative to the listed standard(s).

COMMON CORE STATE STANDARDS FOR MATHEMATICAL PRACTICE

The Common Core State Standards for Mathematical Practice are central to purpose and structure of the activities in *Advanced Common Core Math Explorations: Measurement and Polygons*. Below, we outline the ways in which the activities are built around these standards, providing a few specific examples for purposes of illustration.

1. **Make sense of problems and persevere in solving them.** All of the explorations in the *Advanced Common Core Math Explorations: Measurement and Polygons* book engage students in understanding and solving problems. The process begins when you introduce the activity to your students and have a discussion in which everyone works together to clarify the meaning of the question and think about how to begin. Throughout each exploration, students are involved in devising problem-solving strategies and making and testing conjectures to guide their decisions and evaluate their progress as they work. They discover and describe relationships between geometric drawings, numeric patterns, formulas, and graphs. To promote perseverance, the activities have a high level of cognitive demand, and there is support for the teacher and student in the form of motivation strategies, a tiered structure for the explorations, and suggestions for facilitating mathematical conversation.

TABLE 1

Alignment With Common Core State Standards for Mathematical Content

Exploration	Standard	Connections	Extending the Core Learning
1. Polygon Perambulations	4.MD.C 7.G.B.5 8.G.A.5		Use knowledge of angle relationships to discover and justify patterns and formulas involving interior and exterior angles of convex and concave polygons.
2. Impossible Polygons	4.G.A 5.G.B 7.G.A.2		Analyze and describe relationships between properties of two-dimensional figures by solving challenging problems, some of which have no solution.
3. Starstruck!	7.G.B.5 7.EE.A.2	5.OA.A 5.OA.B 5.G.A	Analyze patterns of angle relationships within complex designs. Describe the patterns using algebraic expressions. Extend the patterns, and make and test predictions about them.
4. Geoboard Squares	6.G.A.1 7.G.B.6	6.EE.A 6.G.A.3	Observe patterns that emerge when creating right angles on a grid. Use them to develop and compare strategies for finding areas of squares whose sides are not parallel to lines of the grid. Analyze and extend the patterns.
5. Creating Area Formulas	6.G.A.1 7.G.B.4 7.G.B.6		Develop and justify area formulas for parallelograms, triangles, trapezoids, kites, and circles. Explore the base/height relationship in depth, and use the ideas to explain why certain formulas work even in "extreme" cases.
6. A New Slant on Measurement	6.G.A.1 7.G.B.6 8.G.B	7.EE.A	Discover the Pythagorean Theorem by analyzing patterns in areas of squares. Verify the discovery algebraically and extend it to solve a problem in three dimensions.
7. Ladders and Saws	5.G.B 7.G.B.5 8.G.A.5	6.G.A.1 8.G.A.2	Apply relationships between parallel lines and angle measures within complex designs to improve geometric visualization skills, develop complex chains of reasoning, and solve challenging problems.
8. Designing Nets	6.G.A.2 6.G.A.4 7.G.B.6 8.G.C.9	8.G.C.7	Understand why the formula $V = B \cdot b$ applies to all prisms. Develop and apply formulas for volume and surface area of pyramids and cones by designing nets and building the shapes.
9. Measuring Oceans	8.G.C.9 8.EE.A.3 8.EE.A.4	7.G.A.3 7.G.B.4 7.EE.B.3 8.G.B.7	Use the formulas for the volume and surface area of a sphere to solve open-ended problems involving order of magnitude estimates of very large quantities. Discover and explain connections between the formulas.

2. **Reason abstractly and quantitatively.** The activities in this book provide students with frequent opportunities to analyze relationships between mathematical concepts and quantities. For example, in the second exploration, Impossible Polygons, students move back and forth between the abstract properties of two-dimensional figures and concrete representations of them as they create shapes having designated sets of properties. In Discovering Area Formulas, they develop formulas for triangles and quadrilaterals by examining relationships between length and area. They are required to think flexibly about the concepts of *base* and *height* as they test their results against more challenging representations of the figures.

3. **Construct viable arguments and critique the reasoning of others.** In these explorations, students use what they have learned in earlier questions and explorations to justify conclusions and explain why certain facts are true. In Ladders and Saws, they begin with a small number of basic assumptions and construct carefully designed chains of logical reasoning to draw conclusions about angle measures within complex drawings. In Polygon Perambulations, students use a variety of diagrams to develop and justify alternate formulas for sums of interior and exterior angles of polygons while analyzing the factors that make the cases different. The Questions and Conversations and Wrap Up features in each exploration provide ongoing support for the teacher to lead discussions in which students compare and critique one another's arguments.

4. **Model with mathematics.** In the exploration, Designing Nets, students create models to build pyramids and cones under a set of constraints that they help to devise. Based on the success of their results, they go back and refine the models as necessary. Afterward, they apply what they have learned to solve real-world problems and develop their own formulas for volume and surface area. In Measuring Oceans, students apply their knowledge of surface area and volume formulas for spheres to develop models for approximating volumes of large bodies of water. They observe the effects of making approximations at different stages of the processes, and they compare models in terms of their effectiveness and efficiency.

5. **Use appropriate tools strategically.** Throughout these explorations, students make choices about using compasses, protractors, rulers, graph paper, calculators, and visual models to solve problems. For example, in Polygon Perambulations, a set of tools is made available to construct regular polygons. Students can discuss the ways in which the choice of tools interacts with the strategies they use to solve the problems. In a number of the explorations, students are encouraged to use dynamic geometry software as an option for exploring the ways in which quantities and properties change (or remain the same) as geometric figures are transformed in different ways.

6. **Attend to precision.** Clear and thorough mathematical communication both during discussion and in written work is a goal for all of the explo-

rations in this book. Students are expected to pay close attention to the precise meaning of all geometric terms as they explain their thinking. This is especially apparent in activities such as Impossible Polygons in which students explore the role of definitions in mathematics. Students' abilities to use drawings to communicate strategies for finding areas and counting solutions are developed in Geoboard Squares. Teachers are provided with support for leading discussions that develop students' communication skills. In the e-booklet accompanying the series, there is a section devoted to helping students understand why it is important to communicate clearly and precisely and how to do so effectively.

7. **Look for and make use of structure.** Pattern and structure are central components of *Advanced Common Core Math Explorations: Measurement and Polygons*. For example, in the exploration, A New Slant on Measurement, students gather data and search for patterns in the areas of "tilted" squares on a grid in order to develop an understanding of the Pythagorean Theorem. In Starstruck, they use polygons to create star-shaped designs. They discover patterns in their reasoning processes that lead to formulas for the measures of the angles at the tips of the stars.

8. **Look for and express regularity in repeated reasoning.** In *Advanced Common Core Math Explorations: Measurement and Polygons*, students regularly engage in processes that display regularity. They use this predictability to find more efficient procedures, develop formulas, and probe connections between concepts. For example, in Geoboard Squares, students use visual models to describe the process of finding the areas of squares whose sides are not parallel to the lines of a grid. By noticing similarities in the way they decompose the squares, they begin to predict the appearance of the decompositions and as a result, they become more efficient in carrying out the process.

Exploration 1

Polygon Perambulations

INTRODUCTION

Materials

- » Ruler, protractor, and compass
- » String or removable floor tape (optional—to create large polygons for students to walk around)

Prior Knowledge

- » Know the definitions of a *regular polygon* and an *interior angle* of a polygon.
- » Know the names of polygons with 10 or fewer sides (triangle, quadrilateral, pentagon, hexagon, heptagon, octagon, nonagon, and decagon).
- » Know that the sum of the measures of the interior angles of a triangle is always 180°.
- » Know how to use a protractor to measure and draw angles.
- » Know that *concave* polygons have one or more "indentations," and *convex* polygons do not.

  Note. This activity works best *before* students know a formula or pattern for finding sums of the interior and exterior angles of a general polygon.

Learning Goals

- » Learn the meanings of exterior and central angles.
- » Visualize relationships between angles in polygons.
- » Develop and justify formulas for sums of interior and exterior angles of a polygon.
- » Analyze numeric patterns and describe them verbally and algebraically.
- » Communicate complex mathematical ideas clearly.

> **Teacher's Note.** The phrase "sums of interior and exterior angles" could be written more precisely as "sums of the measures of interior and exterior angles," because we are not actually adding the angles themselves. Mathematicians (including the writers of the Common Core standards) occasionally use less precise language in order to make sentences less cumbersome. Encourage students to watch for and discuss examples of this!

» Persist in solving challenging problems.

Launching the Exploration

Motivation and purpose. To students: To *perambulate* means to walk around or roam over a territory. In this exploration, your territory consists of polygons, and many of your perambulations will be mental. Historically, the word *perambulation* refers to a process of determining the legal bounds of a region by walking its perimeter. For some questions, you may find it helpful to create polygons large enough that you can physically walk around them and see them from new "angles"!

Understanding the problem. Begin by ensuring that students have the prior knowledge shown above. Review as needed. Then ask students to think of some places where they see regular pentagons in the real world. Students often name the Pentagon building, the inner part of a star, and the surface pattern on a soccer ball. (Soccer balls are covered with both regular pentagons and hexagons.)

As you review names of polygons with students, mention that polygons with many sides are often named directly with a number. For example, a 13-sided polygon is often simply called a 13-gon. (This saves mathematicians the trouble of creating new names for all of these shapes!) When discussing polygons that may have any number of sides, it is common to use the variable n and call them n-gons.

Ensure that each student has access to a ruler, protractor, and compass as he or she begins work. Tell the students that although all of these tools are available to them, it is up to them to decide whether or when to use them.

STUDENT HANDOUT

You may use a ruler, protractor, and compass to help you solve the problems in this exploration.

Stage 1

1. Draw a regular pentagon as accurately as you can. Make it large enough to cover most of a standard sheet of paper. Describe your methods and your thinking processes thoroughly and clearly. What is the measure of each interior angle in the pentagon? Explain.

2. Explain how to extend your method from Problem #1 to the case of regular polygons with *n* sides (*n*-gons). Use specific examples if it makes your explanation clearer.

Stage 2

3. Use the two drawings to develop two different methods for finding the sum of the interior angles of a convex pentagon. Describe your methods and thinking processes thoroughly and clearly.

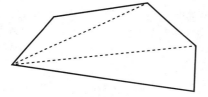 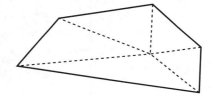

4. Show how to extend one (or both) of your methods to find the interior angle sum for a convex polygon with *n* sides. Describe your methods verbally and as algebraic formulas. Discuss any interesting patterns that you notice or observations that you make.

5. What is the sum of the exterior angles of a convex polygon with *n* sides? Write your answer as a verbal explanation and an algebraic formula. Explain how you found the answer and why it works.

Stage 3

6. Generalize your results for exterior angle sums to the case of concave polygons. Explain your thinking process and justify your conclusions.

7. Generalize your results for interior angle sums to the case of concave polygons. Explain your thinking process and justify your conclusions.

TEACHER'S GUIDE

STAGE 1

Problem #1

1. Draw a regular pentagon as accurately as you can. Make it large enough to cover most of a standard sheet of paper. Describe your methods and thinking processes thoroughly and clearly. What is the measure of each interior angle in the pentagon? Explain.

Questions and Conversations for #1

This section contains ideas for conversations, mainly in the form of questions that students may ask or that you may pose to them. Be sure to allow students to do most of the thinking and talking!

» *Can you think of any calculations that would be helpful?* Think of a way to calculate the size of one or more angles associated with the pentagon.

» *Which part(s) of the pentagon will you draw first?* You might begin by drawing a side first, or you may prefer to begin at the center of the pentagon and work outward.

Teacher's Note. Students' thinking in part (b) often contains an error. Check that they understand where the 72° angle actually is. Many of them believe at first that it is ∠1, which will be an interior angle of the completed pentagon. In order to help them see that ∠1 measures 108° (the other number on the protractor at that point), ask them whether it is acute or obtuse. The 72° angle is actually ∠2. (This will be an *exterior* angle of the completed polygon.)

Solution for #1

The following responses begin with students dividing 360° by 5 to get 72°. They may interpret and apply this result in different ways.

Sample response 1: (a) Draw a line segment ($\overline{AB}$). (b) Align the protractor to the segment at one of its endpoints (B), and make a mark at 72°. Connect the endpoint to this mark with a new line segment ($\overline{BC}$). (c) Make the segment ($\overline{BD}$) the same length as the original. (d) Repeat the process to complete the pentagon.

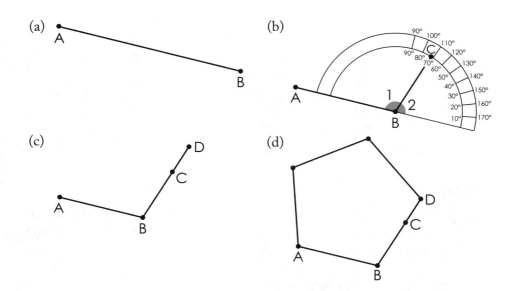

Sample response 2: (a) Choose a point that will be the center of the pentagon. Draw five line segments of equal length radiating from this point, all at 72° angles to their neighbors. (b) Connect the endpoints of these segments to create the pentagon.

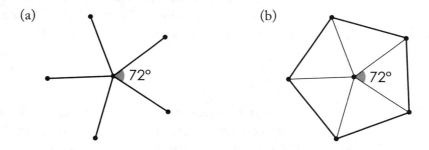

Students who use this strategy may have to work harder to find the measure of an interior angle of the pentagon. They may focus on one triangle in the drawing. Because one of its interior angles has a measure 72°, the measures of the other two angles have a sum of $180° - 72° = 108°$. Since these two angles are congruent, the size of each is $108° \div 2 = 54°$. Each interior angle of the pentagon consists of two of these 54° angles, giving it a measure of $54° \cdot 2 = 108°$.

Sample response 3: (a) Use the compass to draw a circle. Draw a 72° angle using the center of the circle as the vertex. (b) Set the compass to the distance between the two points at which the angle intersects the circle. Use this setting to mark three more points on the circle. (c) Connect the five points on the circle to create the pentagon.

Teacher's Note. The angles having their vertices at the center of the polygon (and circle) are called *central angles*.

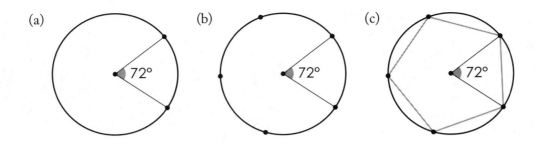

Problem #2

2. Explain how to extend your method from Problem #1 to the case of regular polygons with *n* sides (*n*-gons). Use specific examples if it makes your explanation clearer.

Questions and Conversations for #2

» *Which part(s) of your process from Problem #1 will change? Which will stay the same?* You will divide 360° by a different number to calculate the size of the exterior (or central) angles. The general procedures for drawing the polygons will be the same.

» *How would you describe the appearance of regular polygons that have many sides? What can you say about their interior angles?* These polygons look almost like circles. Their interior angles will be very close to (but still less than) 180°.

Solution for #2

Divide 360° by *n* to find the measure of an exterior angle (or a central angle) for the *n*-gon. Use this angle measure to carry out any of the three strategies shown in the solution to Problem #1. For example, to create a dodecagon (a 12-sided polygon), divide 360° by 12 to get 30°. If you apply this to the second strategy, you will get this type of picture:

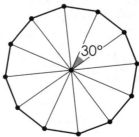

STAGE 2

Problem #3

3. Use the two drawings to develop two different methods for finding the sum of the interior angles of a convex pentagon. Describe your methods and thinking processes thoroughly and clearly.

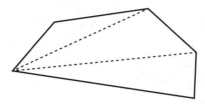 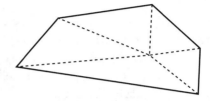

Questions and Conversations for #3

» *Why are the pentagons decomposed into triangles?* You may be able to use the angles in the triangles to solve the problem, because you know the sum of the interior angles of a triangle.

» *What are the important differences between the two pictures?* In the first picture, there are fewer triangles because they were drawn from one vertex of the pentagon. In the second picture, you form the triangles by drawing all possible line segments from a point in the interior of a pentagon to its vertices. Think about how the angles in the triangles relate to the angles in the pentagon.

Solution for #3

Method for the first drawing: Notice that the sum of all of the angles in the three triangles is equal to the sum of the interior angles of the pentagon:

$$180° + 180° + 180° = 180° \times 3 = 540°$$

Method for the second drawing: Notice that the angles surrounding the point inside the pentagon are not part of the interior angles of the pentagon. These angles have a total measure of 360°. To find the sum of the interior angles of the pentagon, add the angles from all five triangles and subtract the 360°:

$$180° \times 5 = 900° \qquad 900° - 360° = 540°$$

Problem #4

4. Show how to extend one (or both) of your methods to find the interior angle sum for a convex polygon with n sides. Describe your methods verbally and as algebraic formulas. Discuss any interesting patterns that you notice or observations that you make.

Questions and Conversations for #4

» *How many triangles are inside an n-gon in the first type of drawing? The second type?* If you are not sure, draw some examples of polygons with varying numbers of sides and count the triangles. Look for patterns in how the number of triangles compares to the number of sides. Your answers to these questions will be algebraic expressions rather than numbers.

Solution for #4

Extended method for the first drawing. Choose one vertex of the polygon and draw a line segment from it to all other vertices except the two closest to it. The number of triangles you form in this process is always two less than the number of sides in the polygon ($n - 2$). Therefore, to find the sum of the interior angles of a polygon (A), subtract two from the number of sides, and multiply the result by 180°. As an algebraic equation:

$$A = 180 \cdot (n - 2)$$

Extended method for the second drawing. Choose a point in the interior of the polygon. Form n triangles by drawing a line segment from every vertex of the polygon to this point. To find the sum of the interior angles of the polygon, first multiply n by 180° in order to find the sum of the interior angles in all of these triangles. Then subtract 360° because the angles surrounding the interior point do not belong to the polygon. As an algebraic equation:

$$A = 180 \cdot n - 360$$

Samples of additional observations and patterns:
» Each time the number of sides increases by 1, the sum of the interior angles increases by 180°.
» In the first type of diagram, you draw a segment from a single vertex to every vertex except itself and the two neighboring vertices. This creates $n - 3$ segments. Because the number of triangles you create is always one greater than the number of segments, there are $n - 2$ triangles.
» The two algebraic equations are *equivalent* (they give the same answers for any value of n) because (1) they both involve multiplying n by 180°, and (2) if you multiply the 2 in the first equation by 180, you get the 360 in the second equation. (Students who are familiar with the *distributive property* may use it to justify this observation.)

Problem #5

5. What is the sum of the exterior angles of a convex polygon with n sides? Write your answer as a verbal explanation and an algebraic formula. Explain how you found the answer and why it works.

Questions and Conversations for #5

» *Based on your experience in Question #1, how would you define an* exterior *angle?* An *exterior angle* of a polygon is the angle between any side of the polygon and an extended adjacent side.

» *How can you begin exploring the exterior angle sums when you have no idea what to expect?* Consider drawing many types of convex polygons, measuring their angles, and calculating the sums. Watch for patterns, and try to explain what causes them.

» *Suppose you take the word "perambulation" literally. What happens if you imagine starting at one vertex and walking all the way around the polygon? What do the exterior angles have to do with this?* Each time you reach a new vertex, you rotate your body to face in the direction of the next side. The exterior angle specifies the angle through which you turn.

> **Teacher's Note.** This question works best after students have spent a substantial amount of time thinking about the problem in other ways. You can use it to further extend students' understanding of exterior angles or to help them get "unstuck" if they are having trouble. Consider using tape or string to create one or more large models of polygons on the floor and asking students to "perambulate them" as they think about this.

» *Why are a central angle and an exterior angle of a regular polygon always congruent?* This is an excellent question for students to think about once they have had a fair amount of experience with exterior angles. Depending on how they solved Question #1, some students may find this to be obvious. Others may devise more complex arguments.

Solution for #5

The sum of the exterior angles of a concave polygon is always 360°. As an algebraic formula:

$$A = 360$$

The fact that the formula does not contain the variable, n, indicates that the sum is constant (i.e., it does not depend on n). Students typically find this result by drawing many polygons and adding the exterior angles.

Sample reason 1. Use a pentagon to illustrate the idea.

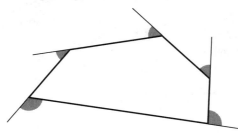

The interior and exterior angle at each vertex combine to make a straight (180°) angle. The sum of the five straight angles is $180° \times 5 = 900°$. If you subtract the sum of the interior angles from this, what remains is the sum of the exterior angles:

$$(180° \times 5) - (180° \times 3) = 900° - 540° = 360°$$

No matter the number of sides, you will always have n groups of 180° for the n straight angles and $n-2$ groups of 180° for the interior angles (see Question #4). When you subtract them, there will always be 2 groups of 180° remaining, which is 360°! You may write this algebraically as:

$$180 \cdot n - 180 \cdot (n-2) = 180 \cdot 2 = 360$$

Most students will not know formal algebraic procedures for simplifying the expression above, but they should be able to explain why the equation makes sense.

Sample reason 2. For students who have "walked the polygon" (See the Questions and Conversations for this problem): Each time you come to a vertex, you rotate through an angle equal to the exterior angle at that point. When you come back to the beginning you will be facing the same direction in which you started. Thus, you will have rotated a total of 360°, which must therefore be the sum of all of the exterior angles.

STAGE 3

Problem #6

6. Generalize your results for exterior angle sums to the case of concave polygons. Explain your thinking process and justify your conclusions.

Teacher's Note. Soon after students have begun thinking about this question, consider suggesting that they view it in terms of "perambulating the polygon." They may visualize the situation more easily if they start by using concave polygons built from right angles (meaning that some of the interior angles will actually be 270°). They are likely to think at first that the exterior angle sum is too large. In this case, you may use the following questions to help them understand what is happening.

Questions and Conversations for #6

» *What does* generalize *mean?* To *generalize* a concept or pattern in mathematics means to extend it so that it applies under a broader set of conditions. In this problem, you are extending something you have learned about convex polygons to the case of concave polygons—and thus, to polygons *in general*. The process involves paying attention to similarities and differences in the two situations.

» *What do you do differently at an "indented" vertex when you walk a polygon?* You rotate in the opposite direction!

» *What happens when you extend the side of a polygon at an "indented" vertex?* It extends into the interior of the polygon.

» *Do you think there is such a thing as a negative angle?* Yes, there is. If you keep track of whether the angle was generated by a clockwise or counterclockwise rotation, you can assign a positive value to one of these and a negative value to the other. (In this case, you usually associate the positive angle with the direction you turn at a vertex that is not indented.)

» *What happens to the angles if you walk the polygon in the opposite direction?* Think about what happens to your rotations when you do this.

Solution for #6

The sum of the exterior angles of a concave polygon is still 360° if you take account of the direction of rotation that produces the angles. If you look at a drawing like this, you are likely to think that the sum is 540°, because you see 6 right angles.

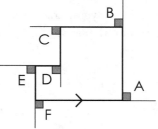

However, if you "perambulate" the polygon beginning at the lower left corner and travel in the direction indicated by the arrow, you will rotate 90° counterclockwise at every vertex except at D, at which point you rotate clockwise because of the indentation there. At this point, you are actually undoing 90° of your progress toward the full 360° counterclockwise rotation needed to close the polygon. Therefore, it makes sense to count the clockwise rotation as a negative 90° angle! The result is:

$$m\angle A + m\angle B + m\angle C + m\angle D + m\angle E + m\angle F = 90° + 90° + 90° + -90° + 90° + 90° = 360°$$

as expected. If you walk in the opposite direction around the polygon, you will count the clockwise rotations as positive, and the counterclockwise rotation as negative. The result will still be 360°.

Teacher's Note. (1) Students may observe that the exterior angle at the "indented" vertex appears to be *inside* the polygon! (2) The *m* before the angle symbol stands for *measure*. It refers to the *size* of the angle rather than the angle itself. This is included because you are actually adding numbers, not the angles themselves. (3) Most students will be comfortable adding negative numbers. However, if they are not, discuss the concept in terms of subtracting positive angles instead.

Students may verify that the sum remains 360° for all concave polygons, regardless of the number of sides or the measures of individual angles. We will illustrate this with one more example.

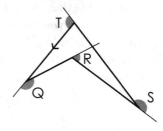

$$m\angle Q + m\angle R + m\angle S + m\angle T = 157° + {-66°} + 164° + 105° = 360°$$

If you begin at the top of the polygon and move in the direction of the arrow, you travel counterclockwise around the polygon. Because the rotation at R is clockwise, it counts as a negative angle.

Teacher's Note. Ask students to apply their reasoning from Problem #2 to the case of concave polygons. They will notice that it is not always easy to decompose them into triangles and interpret the results.

Problem #7

7. Generalize your results for interior angle sums to the case of concave polygons. Explain your thinking process and justify your conclusions.

Questions and Conversations for #7

» *How can you make use of what you learned in Question #6?* Think about what happens when you combine an interior and an exterior angle.

Solution for #7

At each vertex of an n-gon, the exterior angle and the interior angle combine to make a straight angle (180°), resulting in a total measure of $180° \cdot n$ for the entire polygon. Therefore, you can find the sum of the interior angles by subtracting the sum of the exterior angles from $180° \cdot n$. Because you learned in the previous question that the exterior angles always have a sum of 360° (for both convex and concave polygons), the sum of the interior angles is $180 \cdot n - 360$.

Because the argument in the previous paragraph does not depend on whether the polygon is convex or concave, it applies in both cases. In fact, if you had known about sums of exterior angles when you were solving Problem #2, you could have used the same reasoning at that time.

WRAP UP

Share Strategies

Give students an opportunity to compare and contrast their various drawings, observations, conclusions, and justifications. Have them discuss advantages and disadvantages of different approaches. Ask them to pay attention to what they can learn from one drawing as compared to another. Correct any misconceptions that arise.

Summarize

Answer any remaining questions that students have. You may also want to summarize a few key ideas:

» There are at least two formulas for the sum of the interior angles of a polygon: $A = 180 \cdot (n - 2)$ and $A = 180 \cdot n - 360$. These formulas emerge from different types of drawings. They apply to both convex and concave polygons.

» The sum of the exterior angles of every polygon is 360°. You can think of each exterior angle as a rotation that you make when you travel around it. To ensure that the polygon "closes," you must make a total rotation of 360°.

» The formulas for interior and exterior angle sums can be discovered and justified in many different ways. Every student should be able to understand and explain at least one justification for each.

» It makes sense to talk about negative angles if you think of angles as being produced by rotations.

Further Exploration

Ask students to think of new questions to ask or ways to extend this exploration. Here are some possibilities:

» Explore the relationship between n (the number of sides of a polygon) and A (the sum of the interior angles) by making a table and graph. Do the same for the exterior angle sum. If you know about *linear* relationships, *starting numbers* (y-intercepts) and *rates of change* (slopes), discuss these. How can you find them in the table and the graph? Why does the starting number not appear directly in your table for the interior angle sum?

» Can you find more expressions for interior angle sums of polygons by creating and exploring pictures like these?

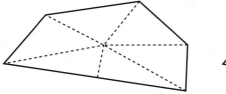 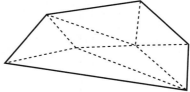

(Sample answers for these pictures: $180 \cdot (n+1) - 540$, $180 \cdot (n+2) - 720$)

Can you find patterns in your expressions? If you know algebraic procedures for rewriting them in equivalent forms, show that your expressions are all equivalent.

» Can you make a triangle having sides of length 3, 4, and 9? What about 6, 7, and 10? What conditions on the side lengths determine whether or not this is possible? Generalize to other types of polygons.

» Generalize the concept of an exterior angle to 3-dimensional shapes (polyhedrons). *Hint*: Think of the exterior angles of a polygon as the amounts by which line segments must bend for the polygon to close. What can you do at each vertex of a polyhedron to help it close? How can you determine an angle at each vertex that measures the extent to which this occurs? What happens when you add these angles together for the entire polyhedron? *Note*. This concept is sometimes referred to as the *angle deficiency* at a vertex of the polyhedron. (Answers: The *angle deficiency* at the vertex of a polyhedron is equal to 360° minus the sum of the angles surrounding that vertex. If the sum of these angles is 360°, the angle deficiency is 0° and the polyhedron looks essentially flat at that vertex. To make it begin to close, you need to remove a polygon in order to decrease this sum. The sum of the angle deficiencies for the entire polyhedron is 720°.)

Exploration 2

Impossible Polygons

Materials

» Ruler and protractor
» Objects for making sides and angles of polygons (recommended)—for example, spaghetti, straws, twist-ties, etc.

Prior Knowledge

» Know that one complete rotation (a full circle) is divided into 360 degrees.
» Classify angles as right, acute, or obtuse.
» Know that the interior angle measures of all triangles have a sum of 180°.
» Identify common types of triangles and quadrilaterals and know their basic properties.

> **Teacher's Note.** It is not expected that students know all of the definitions provided at the end of the student handout.

Learning Goals

» Increase familiarity with definitions and properties of a variety of polygons.
» Understand the role of mathematical definitions in identifying and analyzing geometric figures.
» Analyze relationships between properties of different polygons.
» Understand the distinction between everyday usage of words and mathematical definitions.
» Recognize the value of attending to change when studying geometric figures.
» Communicate complex mathematical ideas clearly.
» Persist in solving challenging problems.

Launching the Exploration

Motivation and purpose. To students: One of the best ways to learn about geometry is to play with shapes. Create them, change them, and watch what happens! Your goal in this exploration is to create geometric figures that have certain combinations of properties. The only catch is that some of them are impossible! You will discover many new things about geometric figures while developing your logical thinking and problem-solving skills. You will also learn about the role of definitions in mathematics.

Understanding the problem. Identify definitions (at the end of the Student Handout) with which students appear to be unfamiliar. Use the Frayer Model (Frayer, Frederick, & Klausmeier, 1969) shown here to explore and analyze them.

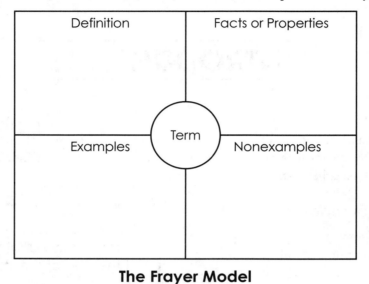

The Frayer Model

For example, students might write the following for the term *concave polygon*.

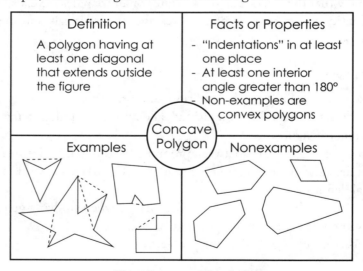

The Frayer Model

Encourage students to collaborate and compare their work. This may lead to additional observations. For example:

> » Some concave polygons have many indentations and many diagonals that extend outside the polygon.
> » Some diagonals may extend only partly outside the polygon.
> » Some diagonals may lie partly along a side.

Students may also notice some important features of mathematical definitions:

> » Definitions do not always directly express the most obvious features of geometric figures.
> » Some definitions take time and effort to understand, partly because of the precision with which they are written.
> » Some definitions are built on others. For example, *irregular* polygons are defined in terms of *regular* polygons, and *concave* polygons are defined using the term *diagonal.* Therefore, to understand some definitions, you must first understand others.
> » Some facts or properties (or combinations of them) may work as alternate definitions. For example, an isosceles triangle may be defined either as a triangle with two congruent sides or with two congruent angles.

The following characteristics of definitions may be less apparent to students, but they are important to discuss and understand:

> » The everyday meanings of some words differ from their mathematical definitions.
> » Your mental image of a geometric figure may not always match its mathematical definition. When there is a conflict between the two, you must rely on the definition.
> » Good definitions generally contain the smallest possible amount of information needed to characterize the term.
> » Definitions are central to mathematical reasoning. The definition is the ultimate source for correctly identifying and reasoning about geometric figures.

Once students have explored and discussed definitions thoroughly, skim the entire exploration with them. Ensure that they understand the instructions in the first paragraph. Emphasize that they should refer to the definitions (a) to resolve any questions they have about geometric figures and (b) to explain their thinking and support any claims they make. If new questions emerge about particular definitions in the course of the activity, return to the Frayer Model and use it to analyze and discuss them.

Teacher's Note. Consider having students use a dynamic geometry application such as GeoGebra (International GeoGebra Institute, 2014) to create and manipulate geometric figures.

STUDENT HANDOUT

Geometric figures cannot have certain combinations of properties. For example, no polygon can ever have exactly four sides but five angles. Your goal in this exploration is to use the definitions provided at the end of the handout to determine whether it is possible to create figures with certain sets of properties. When it is possible, do it. When it is not, explain why. Use the definitions to support your explanations.

Stage 1

1. A decagon with exactly eight vertices

2. An isosceles right triangle

3. A triangle with a pair of parallel sides

4. A scalene triangle with two congruent interior angles

5. A rhombus that is a rectangle

6. A concave quadrilateral

7. An irregular convex heptagon

8. An equilateral obtuse triangle

9. A pentagon with exactly four diagonals

10. A trapezoid whose parallel sides are congruent

Exploration 2: Impossible Polygons

Stage 2

11. A rectangle whose diagonals are perpendicular

12. A parallelogram in which a pair of opposite sides is not congruent

13. A parallelogram in which a pair of opposite angles is not congruent

14. A hexagon in which all sides are congruent, but not all interior angles are congruent

15. A hexagon in which all interior angles are congruent, but not all sides are congruent

16. A kite whose diagonals are not perpendicular

Stage 3

17. A convex pentagon with three acute interior angles

18. A convex pentagon with four acute interior angles

Geometry Definitions

Concave polygon: A polygon having at least one diagonal that extends outside the figure

Congruent figures: Geometric figures that are exactly the same shape and size

Consecutive sides of a polygon: Sides of a polygon that have a common endpoint

Convex polygon: A polygon whose diagonals all lie completely in its interior

Decagon: A 10-sided polygon

Diagonal of a polygon: Any line segment other than a side whose endpoints are two different vertices of a polygon

Equilateral triangle: A triangle having three congruent sides

Heptagon: A seven-sided polygon

Hexagon: A six-sided polygon

Interior of a polygon: The region inside a polygon

Interior angle: An angle inside the polygon (In this exploration, when we speak just of an "angle of a polygon," we mean an interior angle.)

Irregular polygon: A polygon that is not regular

Isosceles triangle: A triangle having two congruent sides

Kite: A quadrilateral having exactly two pairs of congruent consecutive sides

Obtuse triangle: A triangle having an obtuse interior angle

Parallel lines (or segments): Lines that do not intersect (or segments that lie on lines that never intersect)

Parallelogram: A quadrilateral whose opposite sides are parallel

Pentagon: A five-sided polygon

Perpendicular: Intersecting at right angles (applies to lines and line segments)

Polygon: A two-dimensional figure that is made up of line segments joined end to end to make a closed path (the line segments may not cross)

Quadrilateral: A four-sided polygon

Rectangle: A quadrilateral whose interior angles all have a measure of 90°

Regular polygon: A polygon whose sides are all congruent and whose interior angles are all congruent

Rhombus: A quadrilateral having four congruent sides

Right triangle: A triangle having a right interior angle

Scalene triangle: A triangle having no congruent sides

Trapezoid: A quadrilateral having exactly one pair of parallel sides

Vertex (plural: vertices): A corner of a polygon or an angle (a point where the sides of the polygon or the angle meet)

TEACHER'S GUIDE

Geometric figures cannot have certain combinations of properties. For example, no polygon can ever have exactly four sides but five angles. Your goal in this exploration is to use the definitions provided at the end of the handout to determine whether it is possible to create figures with certain sets of properties. When it is possible, do it. When it is not, explain why. Use the definitions to support your explanations.

STAGE 1

Problem #1

1. A decagon with exactly 8 vertices

Questions and Conversations for #1

This section contains ideas for conversations, mainly in the form of questions that students may ask or that you may pose to them. Be sure to allow students to do most of the thinking and talking!

» *When you join line segments end to end without creating a closed figure, how does the total number of endpoints compare to the number of segments?* The total number of endpoints is always one greater than the number of segments. Why?

» *What happens to the number of vertices when you insert a segment that closes the figure?* The number of vertices remains the same (but the number of segments still increases by one).

Solution for #1

Impossible. Every polygon has the same number of vertices as sides. Because a decagon has 10 sides, it has 10 vertices.

Some students may be able to provide more detail. Imagine joining

> **Teacher's Note for the Solutions.** The problems in this exploration have multiple solutions. The answers shown (both drawings and explanations) are samples.
>
> Students will not be able to give rigorous mathematical proofs in their explanations, because they do not yet understand the formal roles of axioms, postulates, definitions, etc. in mathematical reasoning. (They will learn more about these in future courses.) For this and other reasons, their responses are likely to be varied. You should check that students' explanations address the questions, make appropriate use of the definitions, and illustrate a correct understanding of relationships between the properties of polygons.

line segments one at a time without "closing the loop." One segment has two endpoints, so the number of endpoints is one greater than the number of segments.

Each time you join another segment to it, you add one more endpoint so that the number of endpoints remains one greater than the number of segments.

When you insert the segment that closes the figure, it does not add another vertex, because you use the "available" one from the first segment. Therefore, you add a segment without adding a vertex, resulting in the same number of each!

The same reasoning applies no matter how many sides the polygon has.

Problem #2

2. An isosceles right triangle

Questions and Conversations for #2

» *How can it help to make shapes and change them?* Consider beginning with a figure that satisfies one condition, then changing it so that it also fits the other. This is generally easier than trying to satisfy both the "right" and the "isosceles" conditions at the same time.

Solution for #2

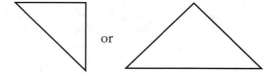

Some students may draw an isosceles triangle and change the angle between the two congruent sides until it is a right angle. Others may create a right angle, make the sides of equal length, and then connect their unattached endpoints to make a triangle.

Problem #3

3. A triangle with a pair of parallel sides

Questions and Conversations for #3

» *What do you know about parallel line segments that you can apply to this problem?* Parallel line segments never intersect, even when extended to form lines.

» *What do you know about triangles that you can apply to this problem?* Because they are polygons, triangles are closed. Every side of a triangle intersects every other side. The interior angle sum for triangles is 180°.

Solution for #3

Impossible.

Sample response 1. In a triangle, each side has a common endpoint with both of the other sides, but parallel sides can never intersect.

Sample response 2. The parallel sides prevent a three-sided figure from being closed. You would need a fourth side to close it.

Sample response 3. Suppose you begin with two parallel segments. When you insert a third segment, it forms two angles that have a sum of 180°. (Students will learn to justify this observation in future courses.) This leaves nothing for the third (interior) angle in the triangle.

Problem #4

4. A scalene triangle with two congruent interior angles

Questions and Conversations for #4

» *If you compare two interior angles in a triangle, can you then make a prediction about the sides opposite them?* Experiment with this by changing the sides (angles) of a triangle and observing the effect on the angles (sides).

Solution for #4

Impossible. If two angles in a triangle are congruent, then the sides opposite those angles are congruent as well, meaning that the triangle is isosceles, not scalene. (Some students may also observe that a longer side is always opposite a larger angle.)

Problem #5

5. A rhombus that is a rectangle

Teacher's Note for #5. Watch for students who make assumptions about geometric figures based on their memory of the general appearance of the shapes rather than on the definitions. For example, some students think of a rhombus as a "diamond." This is not mathematical vocabulary, and it may mislead them into thinking that a square is not a rhombus if it does not appear diamond-shaped to them. Emphasize that the definition is the only completely reliable source for identifying geometric figures.

Solution for #5

Rhombuses and rectangles are quadrilaterals. A rhombus has four congruent sides, and a rectangle has four congruent angles. Therefore, the result must be a square.

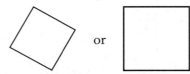

Problem #6

6. A concave quadrilateral

Questions and Conversations for #6

» *Can the diagonal of a polygon be horizontal or vertical?* Yes. The everyday meaning of the word *diagonal* is different than its mathematical definition.

» *What do you know about diagonals of concave figures?* One or more of them lies at least partly outside the figure.

Solution for #6

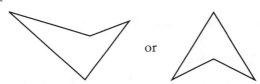

Problem #7

7. An irregular convex heptagon

Solution for #7

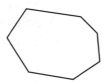

Any seven-sided polygon with no "indentations," whose angles or sides are not all congruent, is a solution.

Problem #8

8. An equilateral obtuse triangle

Questions and Conversations for #8

» What do you know about each interior angle of an equilateral triangle?

Solution for #8

Impossible. An equilateral triangle cannot have an obtuse angle. All of its interior angles must have a measure of 60°.

Problem #9

9. A pentagon with exactly four diagonals

Questions and Conversations for #9

» *Is a diagonal allowed to extend outside a polygon?* Yes. The definition of a diagonal does not require it to lie within the polygon. (Also, see Problem #6.)

» *Suppose you draw a pentagon and all of its diagonals. What happens to the number of diagonals when you drag the vertices to different locations?* The number of diagonals does not change because no new ones are created and none disappear.

Solution for #9

Impossible. All pentagons have exactly five diagonals. (Students can discover this by testing it.)

Problem #10

10. A trapezoid whose parallel sides are congruent

Questions and Conversations for #10

» *What happens if you draw a trapezoid and change it so that its parallel sides become congruent?* Experiment, and watch what happens. Look carefully at the definition of a trapezoid.

Solution for #10

Impossible. This diagram shows an example of what happens. When you make one side (top) congruent to its parallel side (bottom), it forces the other pair of sides to become parallel, making the shape fit the definition of a parallelogram.

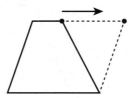

STAGE 2

Problem #11

11. A rectangle whose diagonals are perpendicular

Questions and Conversations for #11

» *What happens to the angles between the diagonals as the rectangle becomes "longer and narrower"?* One pair of angles becomes very small (closer to a measure of 0°) while the other gets closer to 180°.

» *What happens to the angles between the diagonals as the four sides of the rectangle become more nearly the same length?*

Solution for #11

It is a square. (Because a rectangle is a quadrilateral with four right angles, a square is a special type of rectangle.) As the sides of the rectangle become more nearly congruent, the angles between diagonals do the same. When the rectangle becomes a square, these angles become congruent (all 90°), making the diagonals perpendicular.

Problem #12

12. A parallelogram in which a pair of opposite sides is not congruent

Questions and Conversations for #12

» *What happens if you begin with a parallelogram and change it?* Experiment. Make one side of a pair longer than the other and observe the effect.

Solution for #12

Impossible. If you change a parallelogram so that one side is longer or shorter than its opposite side, the other pair of sides is no longer parallel. The dashed segments show this kind of change.

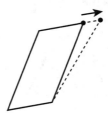

Problem #13

13. A parallelogram in which a pair of opposite angles is not congruent

Questions and Conversations for #13

» *What happens if you begin with a parallelogram and change it?* Experiment. Make one angle larger or smaller than its opposite angle and observe the effect.

Solution for #13

Impossible. If you change a parallelogram so that one angle is larger or smaller than its opposite angle, one or both pairs of opposite sides will no longer be parallel.

Problem #14

14. A hexagon in which all sides are congruent, but not all interior angles are congruent

Questions and Conversations for #14

» *Why might it be helpful to begin with a regular hexagon?* It already satisfies some of the conditions of the problem.

> **Teacher's Note for #14.** Students might find it helpful to experiment with spaghetti, straws, twist-ties, or other hands-on tools to make and manipulate the sides and angles of their hexagons. Dynamic geometry software such as GeoGebra may also be useful.

Solution for #14

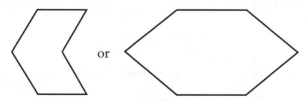

Students may find the first solution by reflecting two consecutive sides of a regular hexagon to its interior. They may find the second one by vertically compressing a regular hexagon.

Problem #15

15. A hexagon in which all interior angles are congruent, but not all sides are congruent

Questions and Conversations for #15

See Questions and Conversations for #14.

Solution for #15

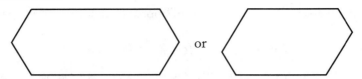

You can obtain these drawings by beginning with a regular hexagon and stretching or shrinking one or more pairs of opposite (parallel) sides so that the sides within each pair remain congruent.

Problem #16

16. A kite whose diagonals are not perpendicular

Questions and Conversations for #16

» *What do you want to draw first?* Experiment with different possibilities. If you draw the diagonals first, you can connect their endpoints to make the kite. Change the angle between the diagonals and watch what happens.

» *Does every pair of intersecting perpendicular line segments form the diagonals of some kite?* No, not unless one line segment *bisects* the other (i.e., splits it into two congruent segments). Does this example help you see why?

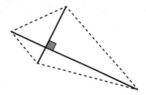

What happens if *both* line segments bisect each other?

» *Can you describe the symmetry of a kite?* One diagonal is a line of symmetry, but the other is not.

» *Does the definition of a kite require that it be a convex polygon? How does this affect your conclusions?* No. According to our definition, the following concave polygon is a kite.

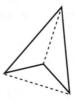

The diagonals (dotted segments) appear to be perpendicular even though they do not intersect.

Solution for #16

Impossible. If you rotate one diagonal of a kite so that the diagonals are not perpendicular, it stretches or compresses the sides by different amounts so that consecutive sides are no longer congruent. (See the dashed segments.)

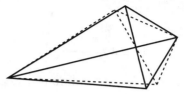

Some students might observe that you can reflect a triangle to create a kite. Choose one side of the triangle, and draw its *altitude** (the dotted segment). When you reflect the triangle over this side, it forms a kite, and the line of reflection (the dashed segment) becomes one diagonal, while the altitude with its reflection becomes the other. Because you can form any kite using this method, the diagonals of a kite will always be perpendicular.

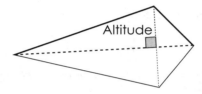

*An *altitude* of a triangle is a line segment perpendicular to one side having the opposite vertex as an endpoint. (Its length is the *height* of the triangle.)

STAGE 3

Problem #17

17. A convex pentagon with three acute interior angles

Questions and Conversations for #17

» *Can you draw a pentagon with three interior right angles?* Yes! Try it. Then think of what to try next.

» *What is true of an interior angle at a vertex where a concave polygon is "indented"?* Its measure is greater than 180°.

Solution for #17

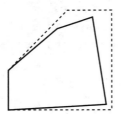

The solid line polygon satisfies the conditions of the problem. Some students may begin with a pentagon having three right angles and adjust them slightly to make them acute. *Note.* Three right angles have a sum of 270°. Because the interior angle sum for a pentagon is 540°, the remaining two angles add to 540° − 270° = 270°, an average of 135° each. When you change the right angles into acute angles, the remaining two angles have to be even larger.

Problem #18

18. A convex pentagon with four acute interior angles

Questions and Conversations for #18

»   *Can you draw a pentagon with four interior right angles?* Extend the ideas from Question #17.

Solution for #18

Impossible. If there are four right angles, their sum is 360°. The fifth interior angle is then 540° − 360° = 180°. Therefore, if the four angles are acute, the fifth angle is greater than 180°. There is an "indentation" at this vertex, meaning that the polygon is not convex.

WRAP UP

Share Strategies

Ask students to share and compare their drawings and explanations. Use this opportunity to correct misconceptions. Students may be surprised at the great variety of ideas and observations.

Summarize

Recap the discussion of the role of definitions in mathematics. (See "Launching the Exploration" from the Introduction to this activity.) Use students' work as a vehicle for illustrating the ideas.

Have a discussion about the usefulness of the strategy of changing the figures and observing the effects. How often was this approach helpful? Could you have used it in situations where you did not? Why is this sometimes better than thinking of shapes as static (unchanging)?

Further Exploration

Ask students to think of ways to continue or extend this exploration. Here are some possibilities:

» Create your own sets of possible and impossible conditions for polygons and test them. Share them with other students and discuss the results!

» Carefully study the definitions in your handout. Some of them could be made more precise. Can you improve any of them?

» Create equivalent definitions for some of the words in the vocabulary list. For example, think of ways to define a parallelogram based on side lengths or angle measures instead of parallel sides. Explain why your definitions are equivalent to the original ones (that is, why they describe exactly the same set of shapes).

» Use tree diagrams or Venn diagrams to illustrate the relationships between categories of polygons.

» Carry out some research to find geometric figures that have alternate definitions. (For example, students may find that some mathematicians define polygons so that sides may intersect at places other than their endpoints or that trapezoids are sometimes said to be quadrilaterals having *at least* one pair of parallel sides.) How do these alternate (nonequivalent) definitions affect the tree diagrams or Venn diagrams discussed above?

Exploration 3

Starstruck!

INTRODUCTION

Materials

- » Ruler, protractor, and compass
- » Graph paper (Problem #4)

Prior Knowledge

- » Complete Exploration 1: Polygon Perambulations through Problem #4 (or have the equivalent knowledge).
- » Review prior knowledge for Exploration 2: Impossible Polygons (as needed).
- » Understand *supplementary* and *vertical angles*.

Learning Goals

- » Apply knowledge of supplementary and vertical angles to solve problems.
- » Make connections between geometric and algebraic reasoning.
- » Represent numeric patterns as algebraic formulas.
- » Create graphs, and use them to analyze patterns of change.
- » Use equivalent algebraic expressions to gain new insight into a problem.
- » Communicate complex mathematical ideas clearly.
- » Persist in solving challenging problems.

Launching the Exploration

Motivation and purpose. To students: In this activity, you will learn to turn geometric thinking strategies into complex algebraic formulas. By analyzing the formulas from multiple perspectives, you expand your insight into the original problem and develop interesting new questions to ask. You also make geometric drawings with beautiful symmetries. When you are finished, you can extend them inward and outward to create even more intricate designs!

Understanding the problem. If necessary, review or discuss definitions of *supplementary* and *vertical* angles.

45

Supplementary angles are pairs of angles that have a sum of 180°.

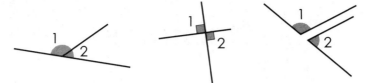

∠1 and ∠2 are supplementary in each of these pictures. You can see this easily in the first picture, because they clearly combine to form a straight angle. However, as the second and third pictures show, supplementary angles may appear in other configurations. It matters only that there are two angles whose sum is 180°.

Vertical angles are opposite angles formed by intersecting lines. The everyday meaning of the word "vertical" is very different than the mathematical definition. Vertical angles need not look "vertical" in the familiar sense of the word.

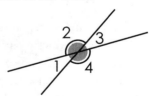

∠1 and ∠3 are vertical angles, as are ∠2 and ∠4. Vertical angles are always congruent. Ask students to apply their understanding of supplementary angles to explain why. (Because ∠1 and ∠3 are both supplementary to the same angle [∠2 or ∠4], they must be congruent to each other.)

If these concepts are new to students, consider asking them to create their own examples and nonexamples of supplementary and vertical angles before they begin the exploration. As students prepare to make their first star, check that they understand the directions to extend each side of the pentagon in both directions. Show them a quick sketch with one or two sides of a pentagon if necessary.

Teacher's Note. This exploration was inspired by an enrichment activity in *Everyday Mathematics* (Bell et al., 2007).

STUDENT HANDOUT

Stage 1

1. Use a ruler, protractor, and/or compass to draw a regular pentagon as accurately as you can. Extend each side of your pentagon in both directions to form a star. Apply your knowledge of angle relationships to find and label the measures of the interior angles in your star without a protractor. Explain your thinking.

2. Repeat the process in Problem #1 for a regular hexagon and a regular octagon. Describe how the stars change in appearance as the number of sides of the polygons increases.

Stage 2

3. Create a formula for the measure of a "star-tip angle" (T) in terms of the number of sides (n) of the regular polygon that was used to construct it. Explain how you found the formula.

4. Use your formula to find star-tip angles for stars made from regular polygons with 5, 6, 7, 8, 9, 10, 11, and 12 sides. Use your data to create a table and a graph. Study your table and graph, and make as many observations as you can. If possible, relate your observations to the appearances of the polygons and stars.

5. What happens when you try to create a star using a square or an equilateral triangle? What happens when you substitute 3 and 4 for n in your formula? What connections do you see between your answers to these two questions?

Stage 3

6. Under what conditions is it possible to create stars from irregular polygons? Explore! Make as many observations as you can about the relationships between the angles in the polygons and the stars. Make your explanations as clear as possible by illustrating your ideas with drawings.

TEACHER'S GUIDE

STAGE 1

Problem #1

1. Use a ruler, protractor, and/or compass to draw a regular pentagon as accurately as you can. Extend each side of your pentagon in both directions to form a star. Apply your knowledge of angle relationships to find and label the measures of the interior angles in your star without a protractor. Explain your thinking.

Questions and Conversations for #1

This section contains ideas for conversations, mainly in the form of questions that students may ask or that you may pose to them. Be sure to allow students to do most of the thinking and talking!

» *How large should your drawing be?* It will be easier to draw, label, and read if it fills most of a standard sheet of paper.

» *How far should you extend the sides of the pentagon?* Extend them until they intersect with other extended sides.

» *What measurements should you use to create the pentagon?* Apply your knowledge of the interior (or exterior) angles of a pentagon.

» *How can you find the interior angles in the star without measuring them?* Apply your knowledge of supplementary angles, vertical angles, and interior angle sums of triangles.

» *Is it necessary to label every angle?* No. Feel free to use your knowledge of the symmetries in the pentagon and star. Because there are five identical sets of angles, it makes sense to label just one set.

» *How many interior angles does the star have?* Ten. Five of them are at the tips of the star. The others are greater than 180°. (Note. Angles larger than 180° are sometimes referred to as *reflex* angles.)

» *What type of polygon is the star (based upon its number of sides)?* The star is a decagon because it has 10 sides.

» *What is the interior angle sum of a decagon?* The interior angle sum of a decagon is $180 \cdot (10 - 2) = 180 \cdot 8 = 1440°$. Check that your results are consistent with this value.

Solution for #1

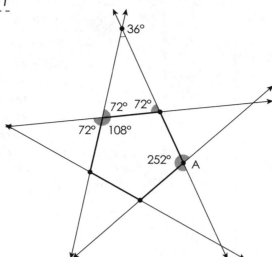

Students are likely to create the pentagon using their knowledge that the interior angles of a regular pentagon have a measure of 108°, although they may use their knowledge of exterior or central angles instead. All sides of the pentagon should be congruent.

The angles at the bases of the isosceles triangles surrounding the pentagon have measures of 72°. The angles at the tips of the star are 36°. The measures of the remaining interior angles of the star at the "indentations" are 252°.

Sample explanation. Students usually notice that the angles at the bases of the isosceles triangles are supplementary to the interior angles of the pentagon, thus calculating $180° - 108° = 72°$. They find the 36° angle at the tip of the star by applying their knowledge of interior angle sums of triangles: $180° - 72° - 72° = 36°$. To determine the reflex angles at the indentations of the star, they add the three angles that comprise them: $108° + 72° + 72° = 252°$.

Be prepared for a variety of responses from students. Although the previous paragraph outlines the most common approach, students are creative! For example, they may combine the fact that the four angles surrounding each vertex of the pentagon must add to 360° with the knowledge that vertical angles are congruent to show that the base angles have measures of 72°.

Problem #2

2. Repeat the process in Problem #1 for a regular hexagon and a regular octagon. Describe how the stars change in appearance as the number of sides of the polygons increases.

> **Teacher's Note.** On this problem and the next, students may feel free to check their predictions of the star angles by measuring them. If the results are not consistent, they should track down the source of the error.

Teacher's Note. The questions for Problem #1 apply to Problem #2 as well. Encourage students to check their predictions by measuring or verifying that the interior angle sums of the stars are consistent with the formula $A = 180 \cdot (n-2)$.

Questions and Conversations for #2

» *What happens to the various angles as the number of sides increases?* They change in different but predictable ways. For example, the star-tip angles become larger, while the angles at the bases of the isosceles triangles become smaller. Look for patterns.

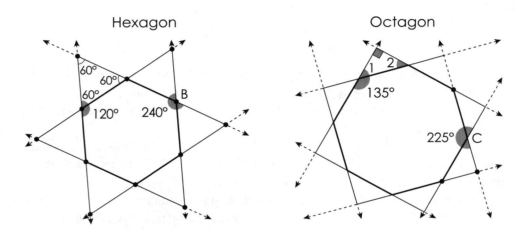

Hexagon | Octagon

What happens as the number of sides becomes very large? If you have difficulty visualizing it, draw or sketch some examples.

Solution for #2

Students may use the same strategies as in Problem #1.

Note. Angles 1 and 2 in the octagon picture have measures of 45°.

Sample observations.

» As the number of sides increases, the base angles of the isosceles triangles become smaller, the star-tip angles become larger, and the reflex angles at the indentations become smaller.

» As the number of sides increases, the tips of the stars move closer to the polygon.

» The star-tip angles and the angles at the indentations get closer and closer to 180° as the number of sides increases. The star-tip angles will never be greater than or equal to 180°. The angles at the indentations will never be less than or equal to 180°.

» As the number of sides becomes very large, both the polygon and its star look more and more like a circle (the *same* circle!).

STAGE 2

Problem #3

3. Create a formula for the measure of a "star-tip angle" (*T*) in terms of the number of sides (*n*) of the regular polygon that was used to construct it. Explain how you found the formula.

Questions and Conversations for #3

» What is an algebraic expression for a single interior angle of a regular *n*-gon?

$$\frac{180 \cdot (n-2)}{n}$$

The numerator represents the sum of the interior angles. When you divide this by *n*, you obtain the measure of each angle. (Students may use the "÷" symbol instead of a fraction to represent division.)

» *What is the relationship between an interior angle of the n-gon and a base angle of the isosceles triangle?* They are supplementary.

» *What variables should be present in your formula?* Your formula should include only the variables *T* and *n*. It should be of the form "*T* = an algebraic expression containing *n*."

Solution for #3

Sample solution 1. The formula that most students develop is surprisingly complex, but it mirrors the details of their earlier calculation process:

$$T = 180 - 2 \cdot \left(180 - \frac{180 \cdot (n-2)}{n} \right)$$

Some students may insert another set of parentheses to make it clear that you multiply by 2 before carrying out the final subtraction. This is fine, but it is not necessary because the standard order of the operations is to multiply before subtracting.

The expression, $\frac{180 \cdot (n-2)}{n}$, at the right end of the formula represents the measure of each interior angle (*A*) in the polygon. When you subtract this expression from 180°, you obtain a base angle of one of the isosceles triangles. You multiply this by 2 because there are two of these base angles. Finally, you subtract this result from 180° to obtain the star-tip angle.

Teacher's Note. Students often write the formula in terms of the interior angle, A:

$$T = 180 - 2 \cdot (180 - A).$$

This does not completely satisfy the conditions of the question, which calls for an expression in terms of n, not A. However, it does show progress in students' thinking.

Consider taking this opportunity to help them understand that they can obtain the answer by substituting $\frac{180 \cdot (n-2)}{n}$ for A in their expression.

(This is an example of *function composition*, which students will learn more about in later algebra courses.)

Sample solution 2.

The following answer is based on the formula $T = 2 \cdot A - 180$ (where again you substitute $\frac{180 \cdot (n-2)}{n}$ for A):

$$T = 2 \cdot \left(\frac{180 \cdot (n-2)}{n} \right) - 180.$$

Although this is a simpler formula than the previous one, fewer students discover it, probably because the reasoning behind it is more complex.

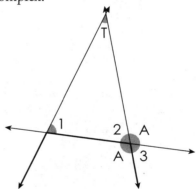

Students may begin by applying their knowledge of supplementary and vertical angles, observing that $\angle 1$, $\angle 2$, and $\angle 3$ are congruent. Because the interior angle sum for every triangle is 180°, they need to determine what to add to the sum of angles 1 and 2 to get 180°:

$$m\angle 1 + m\angle 2 + ? = 180°.$$

However, angles 2 and 3 have the same sum as angles 1 and 2, and "?" is the value of T:

$$m\angle 2 + m\angle 3 + T = 180°.$$

The picture shows that if you add A twice to $m\angle 2 + m\angle 3$, you obtain 360° (the sum of all four angles surrounding the vertex). Because this is 180° too large, you compensate by subtracting 180°. In summary, you add A twice and subtract 180. Therefore, $T = 2 \cdot A - 180$. (Some students may have used this thinking process to calculate the star-tip angles in the first two questions.)

It is exciting to watch students carry out this level of algebraic reasoning before having learned procedures for manipulating algebraic expressions!

Other solutions: $T = \frac{180 \cdot (n-4)}{n}$, $T = \frac{180 \cdot n - 720}{n}$, or $T = 180 - \frac{720}{n}$.

Students who obtain these solutions are not usually thinking about the geometric drawings. Instead they are focusing on the numbers, looking for patterns in pairs of values for n and T, and using trial and error to discover formulas that fit them. Notice that the final formula involves only two calculations!

Problem #4

4. Use your formula to find star-tip angles for stars made from regular polygons with 5, 6, 7, 8, 9, 10, 11, and 12 sides. Use your data to create a table and a graph. Study your table and graph, and make as many observations as you can. If possible, relate your observations to the appearances of the polygons and stars.

Questions and Conversations for #4

» *Does it matter which variable you assign to each axis?* Because you are using n to calculate T, n is the input, and T is the output. The input should be assigned to the horizontal axis.

» *Does it make sense to "connect the dots" on your graph?* Probably not. There are no polygons with fractional numbers of sides. (You sometimes connect dots to make a graph easier to read, but that is not necessary in this case.)

» *What should you look for in your table and graph?* Consider general appearance, predictions about how they would look if you continued them, rates of change, number patterns, connections to formulas, etc.

» *Does the relationship between n and T have a constant rate of change? How can you answer this question using the table? The graph?* No, there is not a constant rate of change. How would you describe the way the value of T does change?

» *For students who have completed the graph: What will the graph look like if you extend it to n = 100 or n = 1000 ?* Think about the vertical scale. How large will the T values be?

» *For students who have completed the table: What happens when you multiply n by its corresponding value of T? What does this product represent?* Consider adding a third row (or column) to your table that shows the value of $n \cdot T$. Look for patterns.

Solution for #4

Students may use any formula from Problem #3 to calculate the values of T (the measure of the star-tip angle) for the given values of n (the number of sides of the original polygon).

Table:

n	5	6	7	8	9	10	11	12
T	36°	60°	$77\frac{1}{7}°$	90°	100°	108°	$114\frac{6}{11}°$	120°

Graph:

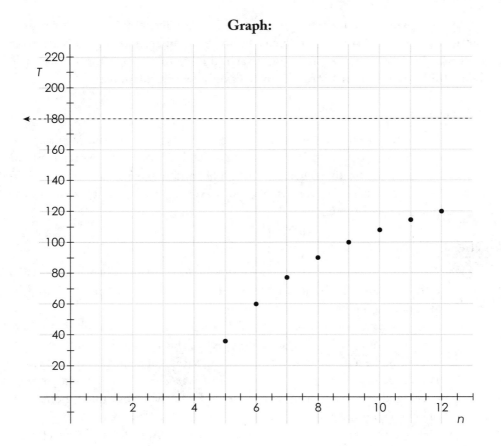

The values of T always increase, although not at a constant rate. In fact, they increase by less and less as n gets larger. This results in a graph that is not a straight line and becomes flatter as you move to the right.

If you were to continue the graph, it would become clear that as n increases, the outputs get closer and closer to $T = 180°$ (the horizontal dotted line) but never quite reach it. (This line is called a *horizontal asymptote* for the graph.) This shows that the star-tip angles become nearly straight as the star begins to look more like a circle. It makes sense that T cannot be greater than 180°, because, if it were, there would be an "indentation" at the star tip. (If students have discovered the formula $T = 180 - \dfrac{720}{n}$, ask them to think about how it shows that T is slightly less than 180° when n is very large.)

Problem #5

5. What happens when you try to create a star using a square or an equilateral triangle? What happens when you substitute 3 and 4 for n into your formula? What connections do you see between your answers to these two questions?

Questions and Conversations for #5

» *How are star-tip angles formed?* They result from the intersections of the lines that extend the sides of the polygon. Do these lines intersect when you begin with squares and triangles?

Solution for #5

It is not possible to create a star from a square or an equilateral triangle. In the case of the square, the pairs of extended lines are parallel. Therefore, they never intersect to create the angles that would form the tips of a star. In the case of the equilateral triangle, they intersect at the vertices of the triangle. Thus, they "cross over" each other as they extend beyond the triangle.

Teacher's Note. Some students find it interesting to explore the product of n and T, which represents the sum of the star-tip angles in a star. The results are multiples of 180°!

n	5	6	7	8	9	10	11	12
T	36°	60°	$77\frac{1}{7}°$	90°	100°	108°	$114\frac{6}{11}°$	120°
$n \cdot T$	180°	360°	540°	720°	900°	1080°	1260°	1440°

Therefore, the sum of the star-tip angles of a star created from a polygon with n sides is equal to the sum of the interior angles of a polygon with $n-2$ sides. In fact, the sum of the star tip angles is $180 \cdot (n-4)$. Notice the connection to the formula $T = \dfrac{180 \cdot (n-4)}{n}$.

(See the solutions to Problem #3.)

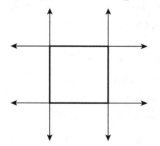 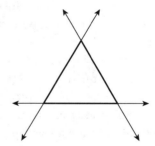

When you substitute $n = 4$ and $n = 3$ into any of the formulas for T, you obtain the answers 0° and −60° respectively. The fact that the measures of the star-tip angles are not positive numbers indicates that creating a star is impossible. Some students may observe that the 0° angle suggests parallel lines, while the negative value, −60°, represents extended lines crossing over one another.

STAGE 3

Problem #6

6. Under what conditions is it possible to create stars from irregular polygons? Explore! Make as many observations as you can about the relationships between the angles in the polygons and the stars. Make your explanations as clear as possible by illustrating your ideas with drawings.

Solution for #6

Sample observations:

» The star-tip angle is given by the formula $T = m\angle a + m\angle b - 180$, where $\angle a$ and $\angle b$ are in the positions shown below. (Notice the similarity to the earlier formula, $T = 2 \cdot A - 180$.)

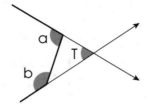

» It is possible to create a star from a polygon if every pair of consecutive (neighboring) angles in the polygon has a sum greater than 180°. This condition is consistent with the first observation, because it ensures that T is always positive.

» The previous observation implies that is never possible to create a star from a triangle or quadrilateral. (Can you see why?)

» If $m\angle a + m\angle b = 180°$, the associated lines are parallel. If the sum is less than 180°, the lines diverge as they leave the polygon. In both cases, the lines fail to create a star-tip.

$$m\angle a + m\angle b = 180° \qquad m\angle a + m\angle b < 180°$$

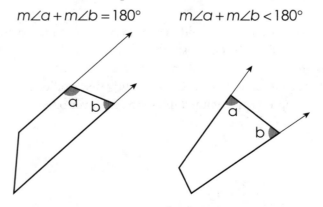

The sum of the star-tip angles is always $180 \cdot (n - 4)$, whether the original polygon is regular or not (when it is possible to create the star).

WRAP UP

Share Strategies

Give students an opportunity to compare their observations, strategies, and formulas. Discuss advantages and disadvantages of each.

Summarize

Answer any remaining questions that students have. Summarize and expand upon a few key ideas:

» You can use supplementary and vertical angles to analyze complex drawings.

» As the number of sides increases, (1) the interior angles of polygons become larger, (2) the base angles of the isosceles triangles become smaller, and (3) the star-tip angles become closer to 180°. As a result, both the star and the polygon come to look more like a circle.

» Different reasoning processes may lead to different expressions for a pattern. These expressions are *equivalent* in the sense that they give the same answers (for the same inputs). Equivalent algebraic expressions may give you different insights into a problem situation. For example

- The expression $180 - 2\left(180 - \dfrac{180 \cdot (n-2)}{n}\right)$ represents each step of the most common procedure for calculating a star-tip angle.

- The expression $2 \cdot \dfrac{180 \cdot (n-2)}{n} - 180$ is simpler than the previous one, even though it is related to a more complex reasoning process.

- The expression $\dfrac{180 \cdot (n-4)}{n}$ calls attention to the sum of the star-tip angles, $180 \cdot (n-4)$, and suggests a connection to the interior angle sums of polygons with fewer sides.

- The expression $180 - \dfrac{720}{n}$ is very efficient. It also shows why the star-tip angles approach but never reach 180° as n increases.

» In algebra courses, you learn procedures to transform algebraic expression into equivalent ones. You could use these procedures, for example, to change the first (complex) expression into the last (simple) one.

Further Exploration

Ask students to think of new questions to ask or ways to extend this exploration. Here are some possibilities:

» Find an algebraic expression for the reflex angles in the stars. Check that your result is consistent with the interior angle sum formula for polygons.

» If you know procedures for writing algebraic expressions in equivalent forms, use them to translate between the various expressions for the star-tip angles.

» Expand your drawings inward and outward. Use them to make new observations and generate new questions. For advanced students who know properties of square roots, ratios, areas of polygons, etc., explore ratios of lengths and areas in your drawings.

Exploration 4

Geoboard Squares

INTRODUCTION

Materials

- » Dot paper or graph paper
- » Geoboards and rubber bands (optional)

Prior Knowledge

- » Know how to plot points on a coordinate grid.
- » Know the meaning of *area*. Find areas of rectangles on a grid.

> **Teacher's Note.** This exploration works best *before* students have learned to use the Pythagorean Theorem to find lengths of line segments.

Learning Goals

- » Explore perpendicular segments in a coordinate grid.
- » Develop and apply strategies for finding areas of polygons on a grid.
- » Collect and organize data.
- » Analyze and extend numeric and geometric patterns.
- » Generate algebraic expressions to describe numeric patterns (Stage 3).
- » Communicate complex mathematical ideas clearly.
- » Persist in solving challenging problems.

Launching the Exploration

Motivation and purpose. To students: Often, there is more to a math problem than meets the eye! This investigation is a good example. As you challenge yourself to visualize squares on geoboards, you will deepen your understanding of area and perpendicular lines. But keep your eyes open for the unexpected. Right when you think you have finished the problem, you may discover that you have just begun!

Understanding the problem. Introduce students to *geoboards* if they are not familiar with them. Geoboards have pegs between which you stretch rubber bands to create geometric figures. If you have geoboards available in your classroom, students may enjoy using them, especially at the beginning of the exploration. In my experience, they eventually come to prefer using dot paper or graph paper because it is more efficient.

Ask students what *area* is. They often respond by giving verbal descriptions of formulas such as "length times width" or "pi r-squared." Help them understand that although formulas enable you to calculate areas in certain cases, they are not what area *is*. The *area* of a 2-dimensional (flat) region is a number that describes its size. You can find areas without using formulas by choosing a unit region (usually a square), and determining the number of copies of it needed to cover the region. If figures are drawn on dot or graph paper, as they will be in this activity, you can count the square units (or partial square units) inside the figure. You can also develop more sophisticated strategies when this counting process is impractical or imprecise.

Look through the exploration with students. The key question is Problem #3, in which they find all of the squares that can be formed on a 6-peg by 6-peg geoboard. Problems #1 and #2 develop some helpful background about area and perpendicular lines. In Problems #4 and #5, they generalize the problem to geoboards of different sizes and look for patterns.

If necessary, remind students how to plot points in a coordinate grid and review some vocabulary for Problem #1. *Perpendicular* lines (rays/segments) are lines that intersect at a 90° angle (or rays/segments that lie on such lines). A *rectangular coordinate system* (*grid*) is shown below. It consists of two perpendicular number lines: the horizontal *x-axis* and the vertical *y-axis*. The two axes intersect at the *origin* and separate the surface (*plane*) into four regions known as *quadrants*, which are traditionally labeled with the Roman numerals I, II, III, and IV.

> **Teacher's Note.** The final page of the student handout provides a template to help students organize their work for Problem #3. They will need two copies in order to show all of the solutions. Consider giving them just one and asking them to let you know if they need another copy. If you believe that students will benefit from organizing the information themselves, there is no need to give them the template at all!

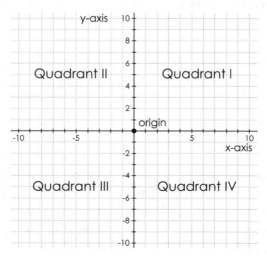

STUDENT HANDOUT

Stage 1

1. In a rectangular coordinate grid, draw a line segment from the origin to the point with the given coordinates. Then, starting at the origin, draw two rays perpendicular to the segment. On each ray, mark and label the coordinates of the point that is the same distance from the origin as is the original point.

    a. (5, 2)   b. (−3, −4)   c. (−1, 4)

    Look closely at the coordinates of each of the three points in your drawings. Describe any patterns that you find, and explain what causes them.

2. Find the areas of the polygons. Show and explain your strategy for each.

    a. 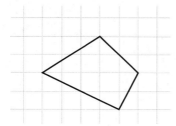   b.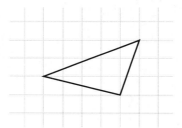

3. How many squares can you make on a 6-peg by 6-peg geoboard? (Every vertex lies on a peg.) Include the following elements in your written solution:
    » Separate your squares into categories of congruent squares. Draw one or two examples from each category.
    » Find the number of squares in each category. Explain how you counted them.
    » Show how to find the area of a square in each category.
    » Find the total number of squares on the geoboard.

Advanced Common Core Math Explorations: Measurement & Polygons © Prufrock Press Inc.

Exploration 4: Geoboard Squares

4. Extend your investigation by counting the squares you can make on square geoboards of different sizes. Make a table of your results. Look for patterns in your tables. Show how to use the patterns to predict the numbers of squares for larger geoboards.

Stage 3

5. Find a formula for the number of squares on an n-peg by n-peg geoboard.

SOLUTION TEMPLATE
FOR PROBLEM #3

Category	Sample Square(s)	Number of Squares	Area Picture	Area (units²) With Calculation

Advanced Common Core Math Explorations: Measurement & Polygons © Prufrock Press Inc.
Permission is granted to photocopy or reproduce this page for single classroom use only.

Exploration 4: Geoboard Squares

TEACHER'S GUIDE

STAGE 1

Problem #1

1. In a rectangular coordinate grid, draw a line segment from the origin to the point with the given coordinates. Then, starting at the origin, draw two rays perpendicular to the segment. On each ray, mark and label the coordinates of the point that is the same distance from the origin as is the original point.

     a. $(5, 2)$ b. $(-3, -4)$ c. $(-1, 4)$

Look closely at the coordinates of each of the three points in your drawings. Describe any patterns that you find, and explain what causes them.

Questions and Conversations for #1

This section contains ideas for conversations, mainly in the form of questions that students may ask or that you may pose to them. Be sure to allow students to do most of the thinking and talking!

» *What if you can't read the exact coordinates of the points on the rays?* Draw your rays more carefully. The required coordinates are integers (whole numbers or their opposites). They should not be hard to read from the grid.

» *What kinds of patterns are you looking for?* Compare the coordinates of the points on the rays to those of the original point. Compare the coordinates of the rays to each other.

Solution for #1

a.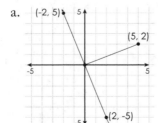
b.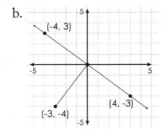
c.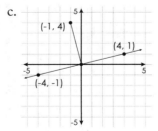

To find the coordinates of the points on the rays, interchange the x and y coordinates of the original point, and then take the opposite of one of them. Look at part (a) to see why this makes sense. For all three points, you start at the origin and travel 5 units in one direction and 2 units at a right angle to this. Therefore, the point stays at the same distance from the origin.

The following diagram illustrates the reason that the rays form 90° angles with the segment. To travel from the origin to the point at $(5, 2)$, you move 5 units to the right and 2 units up. Now imagine rotating the triangle 90° as shown below.

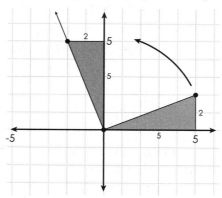

"5 units right and 2 units up" becomes "5 units up and 2 units left." Left/right motion is interchanged with up/down motion. You can check that this is true regardless of the location of the original segment and whether you rotate clockwise or counterclockwise.

Problem #2

2. Find the areas of the polygons. Show and explain your strategy for each.

a.

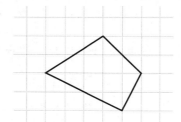

b.

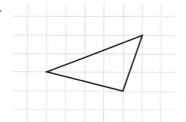

Questions and Conversations for #2

» *How do you label the areas when no specific units are given?* Use the term *square unit* (or *unit²*).

» *Can you decompose the figure into simpler shapes? If so, what types of shapes would be most helpful?* Decomposing the figure into rectangles and right triangles is helpful

Teacher's Note. Some students count squares inside the figures, estimating the sizes of the partial squares. Often, they rearrange partial squares to form whole squares. Allow them to pursue these methods because they reinforce students' understanding of the meaning of area. Afterward, ask them how confident they are in the exactness of their answers. They should have some doubt. If they cannot think of strategies that will increase their confidence, follow up with questions such as the ones below.

because the right angles make it easy to find areas. However, some shapes cannot be decomposed in this way.

» *How can you find the exact area of a right triangle on the grid?* Join it to a copy of itself to form a rectangle. What is the relationship between the area of the rectangle and the area of the triangle?

» *Might it help to surround the figure with a simpler shape?* Consider surrounding it with a rectangle.

Solution for #2

a. 10 units^2 b. 6.5 units^2

There are two common strategies for calculating areas such as these on a grid: (1) decompose the figure into simpler shapes as illustrated in part *a*, and (2) surround and subtract, as shown in part *b*.

a. 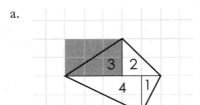   b.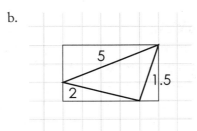

In part *a*, students decompose the quadrilateral into right triangles and find their areas by viewing them as halves of rectangles. For example, the triangle in the upper left of the figure has half the area of the rectangle shown in gray. Because the area of the rectangle is 6 units^2, the area of the triangle is 3 units^2. The total area is $3+2+4+1=10$ units^2.

In part *b*, students surround the figure with a rectangle and subtract the extra area. Because the total area of the rectangle is 15 units^2, and the areas of the extra triangles are 5, 2, and 1.5 units^2, the area of the triangle is $15-5-2-1.5=6.5$ units^2. *Note.* The "surround and subtract" strategy would also work in part *a*. Try it!

Problem #3

3. How many squares can you make on a 6-peg by 6-peg geoboard? (Every vertex lies on a peg.) Include the following elements in your written solution:

» Separate your squares into categories of congruent squares. Draw one or two examples from each category.

» Find the number of squares in each category. Explain how you counted them.

» Show how to find the area of a square in each category.

» Find the total number of squares on the geoboard.

Questions and Conversations for #3

» *How many 1 by 1 squares fit into the top row of the geoboard?* Five. (Be careful not to assume that it is six!) *Note.* A 1 by 1 square has 2 rows and 2 columns of pegs. We describe the size of a square using its length and width instead of the number of pegs.

» *What does "categories of congruent squares" mean, and why are you asked to draw just one or two examples in each category?* Every square within a category must be exactly the same size and shape. You are asked to draw only one or two examples of each in order to save time and to make your diagrams easier to read.

» *Can you think of an efficient way to determine the number of squares in each category without drawing all of them?* There are at least two options: (1) focus on one vertex for each square, or (2) pay close attention to how the squares are arranged in rows and columns.

» *Does Problem #1 give you any ideas for finding more squares?* If it doesn't, you should spend more time thinking about it!

» *How will you know when you've found all of the squares?* Be systematic in your search and in the way you organize your answers on paper.

Solution for #3

There are 105 squares in a 6-peg by 6-peg geoboard. Tables 1 and 2 show examples of the different types of squares, the number of each type, their areas, and a picture that shows a strategy for calculating the areas. (Students may list the categories in any order they like.)

TABLE 1

Category	Sample Square(s)	Number of Squares	Area Picture	Area (units²) With Calculation
1		25		$1 \cdot 1 = 1$
2		16		$2 \cdot 2 = 4$
3		9		$3 \cdot 3 = 9$

Table 1. *continued.*

Category	Sample Square(s)	Number of Squares	Area Picture	Area (units²) With Calculation
4		4		$4 \cdot 4 = 16$
5		1		$5 \cdot 5 = 25$
6		16		$\frac{1}{2} \cdot 4 = 2$
7		4		$2 \cdot 4 = 8$

Teacher's Note. Watch for students who initially have the misconception that the square in Category 1 is the same size as the square in Category 6. These students may need to see that a diagonal joining opposite vertices of a square is longer than the side of the square. That the squares are not congruent is further verified by the fact that they have different areas.

Students tend to find the 55 squares in Categories 1–5 of Table 1 quite quickly. They usually calculate the areas simply by counting the unit squares within them or by multiplying the number of rows by the number of columns. The open dots in each diagram represent the allowable locations of the upper left vertex of each square. This is an efficient method of counting the squares in each category.

Categories 6 and 7 in Table 1 contain 20 additional solutions. It usually takes students a little longer to find these. To find the areas of these squares, students may decompose them into right triangles as shown. As before, the open dots provide an easy way to count the squares in the category.

TABLE 2

Category	Sample Square(s)	Number of Squares	Area Picture	Area (units²) With Calculation
8		$9 + 9 = 18$		$1 + (1 \times 4) = 5$
9		$4 + 4 = 8$		$4 + (1.5 \times 4) = 10$
10		$1 + 1 = 2$		$9 + (2 \times 4) = 17$
11		$1 + 1 = 2$		$1 + (3 \times 4) = 13$

The 30 solutions in Table 2 are generally more difficult to find. Notice that each type of square in Table 2 has a reflected partner. The open dots represent the possible locations of the upper left vertex of one of the squares. There are an equal number of reflected squares in each diagram, although I have not shown the dots for these.

Table 2 illustrates the decomposition strategy for determining the areas of the squares. Students use this approach frequently, but they may also employ the surround and subtract strategy illustrated below.

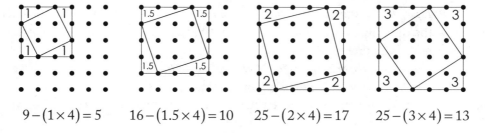

$$9 - (1 \times 4) = 5 \qquad 16 - (1.5 \times 4) = 10 \qquad 25 - (2 \times 4) = 17 \qquad 25 - (3 \times 4) = 13$$

69

As in part (b) of Problem #2, subtracting the areas of the right triangles from the area of the surrounding square produces the area of the "tilted" square. Of course, students may develop other strategies and pictures for calculating these areas!

To obtain the total number of squares on the geoboard, add the number of squares in each category:

$$25 + 16 + 9 + 4 + 1 + 16 + 4 + 18 + 8 + 2 + 2 = 105$$

STAGE 2

Problem #4

4. Extend your investigation by counting the squares you can make on square geoboards of different sizes. Make a table of your results. Look for patterns in your tables. Show how to use the patterns to predict the numbers of squares for larger geoboards.

Questions and Conversations for #4

» *What does a 1-peg by 1-peg geoboard look like?* It has just a single peg! (No square can be formed.)

» *How can you ensure that you don't miss any of the "tilted" squares?* Be systematic. Find a way to label the sides. For example, consider using the horizontal and vertical number of spaces you move in the process of traveling from one endpoint of a side to the other. You could call these the "tilt numbers" of the square. The following picture shows a square whose sides have tilt numbers of 2 and 4.

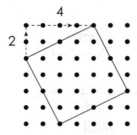

» *How do you start if the table doesn't appear to show any patterns?* Consider the differences between the numbers of squares. If you don't see a pattern, repeat the process as needed.

» *Once you make predictions for larger geoboards, how can you be confident in your results?* Test your predictions by finding all of the squares and counting them. (This gets easier once you've done it a number of times, because you become familiar with the patterns.)

Solution for #4

The black text in the first two rows of the table shows the results for the six smallest boards. The final two rows show patterns of change. The number of squares increases by 1, 5, 14, 30, and 55. These "First Differences," in turn, increase by 4, 9, 16, and 25. These are the "Second Differences," and they appear to be the square numbers!

Number of Pegs Per Side	1	2	3	4	5	6	7	8
Number of Squares	0	1	6	20	50	105	196	336
First Difference	+1	+5	+14	+30	+55	+91	+140	
Second Difference		+4	+9	+16	+25	+36	+49	+64

Given the fact that square numbers have surfaced in many places in this problem, it is reasonable to suppose that this pattern continues—and, in fact, it does! The next entries in the final row are 36 and 49. You can use these to find the next two first differences:

$$55 + 36 = 91 \text{ and } 91 + 49 = 140$$

Therefore:

A geoboard with 7 pegs per side holds $105 + 91 = 196$ squares, and a geoboard with 8 pegs per side holds $196 + 140 = 336$ squares. Students may extend these patterns for as long as they like!

STAGE 3

Problem #5

5. Find a formula for the number of squares on an n-peg by n-peg geoboard.

Questions and Conversations for #5

» *What types of operations are likely to be involved?* Because the number of squares on the geoboard changes more and more rapidly, the formula may involve multiplication or powers.

» *What types of numbers are likely to play a role in the formula?* Because square numbers appear to be important in the problem, they may also show up somewhere in the formula.

> **Teacher's Note.** This is a very challenging question. I typically have one or two students per year who find a correct formula, and they usually need a few hints along the way. Most students can learn from experimenting, though—and they become intensely curious about the solution when they are unable to find one!

Teacher's Note. If students look at their geoboard data from Problem #3, they will see that the process of adding square numbers appears to be very important in this problem. Allow them plenty of time to search for these expressions. Some will find them, and those who do not may still display great ingenuity in their search!

Once students know both expressions, they tend to get a feel for the type of answer they will be looking for in Problem #5. For example, they may expect a fraction with a denominator greater than 6 having factors of 2 and 3. They may also suspect that the numerator will be a product of simple linear expressions involving n, or a sum or difference of small powers of n.

» *Can you find an expression to calculate* $1+2+3+...+n$? Consider exploring the following type of picture.

What happens if you make a copy of the "stairs," turn it upside-down, and join it to the original diagram? You can evaluate the sum with the expression $\frac{n(n+1)}{2}$. Try it! *Note.* Students may find other versions of the expression, such as $\frac{1}{2}(n^2+n)$.

» Can you find an expression to calculate $1^2+2^2+3^2+...+n^2$? Some students may consider creating and exploring a 3-dimensional version of the staircase diagram above. You can evaluate the sum using the expression $\frac{n(n+1)(2n+1)}{6}$.

Solution for #5

The expression may be written in many forms. Three possibilities are:

$$\frac{n^2(n-1)(n+1)}{12}, \frac{n^2(n^2-1)}{12}, \text{and } \frac{n^4-n^2}{12}.$$

Teacher's Note. Some students may predict the existence of the factor $n-1$ in the first expression (or n^2-1 in the second) by noticing that it will produce an output of 0 when $n=1$. Others may be curious about the n^4 in the final expression. Consider asking them to extend the table in Problem #4 to show third and fourth differences. The fact that the fourth differences are constant is connected to the fact that the greatest exponent in the final expression is 4.

WRAP UP

Share Strategies

Give students an opportunity to share results and strategies for finding and counting squares, determining areas, and organizing their work. If they completed Problems #4 or #5, ask them to compare patterns and algebraic expressions. If students arrived at different expressions in Problem #5, ask them if it is possible for more than one expression to be correct. Encourage them to test the expressions by extending their tables further and checking that they continue to give the correct number of squares. Some students may be able to use algebraic properties (such as the distributive property) to prove that different expressions are equivalent.

Summarize

Answer any remaining questions that students have. You may also want to summarize and expand on a few key ideas:

» If a point in a rectangular coordinate grid is rotated 90° about the origin, its coordinates are reversed, except that one of them becomes the opposite of its initial value. This can be expressed symbolically as $(x, y) \rightarrow (y, -x)$ or $(x, y) \rightarrow (-y, x)$. You may think of this as interchanging up/down and left/right motions when traveling from the origin to the points. (*Note.* In later courses, this will be seen to relate to slopes of perpendicular lines being opposite reciprocals.)

» You can find the area of a right triangle by thinking of it as half of a rectangle.

» You can determine areas on a coordinate grid using (1) decomposition or (2) surround and subtract strategies.

» Squares on grids do not always have vertical and horizontal sides.

» You can often make predictions about complex patterns by studying them closely. It may be helpful to look at differences in the outputs.

Further Exploration

Ask students to think of new questions to ask or ways to extend this exploration. Here are some possibilities:

» Look for patterns in the number of squares in each category (across square geoboards of different sizes). Use the patterns to make predictions.

» Investigate the number of squares on rectangular (nonsquare) geoboards.

» Investigate the number of rectangles on square geoboards.

» Investigate the number of rectangles on rectangular geoboards.

» Investigate the number of triangles on geoboards of different sizes. What about isosceles or equilateral triangles?

» For those with plenty of algebraic knowledge and experience: Prove the formula that you found in Stage 3.

Exploration 5

Creating Area Formulas

```
┌─────────────────────────────────┐
      INTRODUCTION
└─────────────────────────────────┘
```

Materials

» Scissors—for cutting and rearranging geometric figures (recommended)
» Graph paper

Prior Knowledge

» Complete Exploration 2: Impossible Polygons, Stages 1 and 2 (recommended).
» Understand and calculate areas of rectangles and right triangles (for both whole number and fractional measurements).
» Know a formula for the circumference of a circle.
» Vocabulary: *parallel, perpendicular, diagonal, diameter, radius, circumference, names of special quadrilaterals*

Learning Goals

» Use conceptual understanding of area to devise and compare strategies for creating and testing area formulas.
» Learn formulas for areas of parallelograms, triangles, trapezoids, kites, and circles, and explain why they work.
» Understand the terms *base, altitude,* and *height,* and know when it is possible to make choices about which side of a geometric figure represents the base.
» Communicate complex mathematical ideas clearly.
» Persist in solving challenging problems.

Launching the Exploration

Motivation and purpose. To students: Students are often accustomed to finding formulas in a book and memorizing them. But where do the formulas come from? In this exploration, you invent or discover area formulas yourself. By doing this, you gain a deeper and more flexible understanding of them. You also learn to be a *producer,* not just a consumer, of mathematical knowledge!

Understanding the problem. Look through the exploration with students so that they know what to expect. In Stage 1, they develop well-known formulas for geometric figures. In Stages 2 and 3, they discover a less familiar formula and explore others in greater depth.

For Problem #5, review the circumference formula, $c = \pi \cdot d$ or $c = 2 \cdot \pi \cdot r$. Ask students why it makes sense. Have them imagine what happens if they bend a diameter around a circle. How many times will it fit? *No matter the size of the circle*, it always fits approximately 3.14 (or exactly π) times, as illustrated here.

To further prepare students for Problem #5, ask them to think of an expression for half the circumference of a circle. (*Answer:* $\pi \cdot r$.)

Some points to consider:

» Students who need to begin with a more concrete approach for finding formulas will benefit from choosing numerical values for the measurements before working with variables. Alternatively, they could draw the figures on graph paper.

» This exploration is more about *creating* formulas than using them. Many students will benefit from computational practice to develop comfort and fluency with them.

» If some students already know one or more formulas, increase their motivation to explore further by creating a sense of doubt. Show them unusual or unfamiliar examples of the figures such as very "tilted" parallelograms, obtuse triangles, or trapezoids such as the one below. Ask: Are you sure the formula always works—even in cases like these? How do you know?

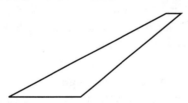

STUDENT HANDOUT

Stage 1

1. Draw a parallelogram like the one below. Show how to find a formula for its area using your knowledge of rectangle area.

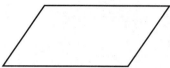

2. Draw a triangle like the one below. Show how to find a formula for its area using your knowledge of rectangle area. Can you use the side lengths as your variables? If not, what could you use instead? Explain.

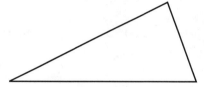

3. Show how to create as many different (noncongruent) polygons as you can by joining two copies of this triangle side to side. Choose one of your drawings and explain how to use it to find a formula for the area of the triangle. Is it the same as the formula from Problem #2?

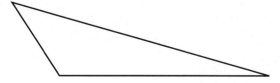

4. Draw a trapezoid like the one below. Apply your knowledge of previously discovered formulas to find a formula for its area, determining the variables that are needed as you work. Explain your thinking.

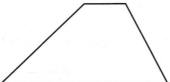

5. Use the first drawing to explain why an expression for the area of a circle appears to be between $3 \cdot r^2$ and $4 \cdot r^2$ (but closer to $3 \cdot r^2$). Use the second drawing to explain why the formula for the area of a circle is $A = \pi \cdot r^2$.

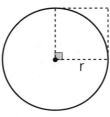 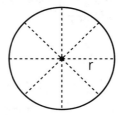

Stage 2

6. Develop a formula for the area of a (convex) kite in terms of the lengths of its two diagonals, d_1 and d_2. Show how you found the formula and why it works. Does your formula apply to concave kites? Explain.

7. Find the area of the triangle on p. 80 as accurately as you can. Carry out the process three times, using a different side as the base each time. Make any necessary additions to the drawing as you work. Report your answers with an appropriate level of precision. Do all three answers agree?

Stage 3

8. Show how to find the exact area of this parallelogram (in units²) without using a formula. Would the formula from Problem #1 still work? Why or why not?

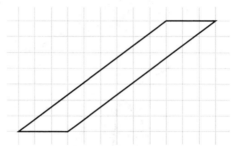

9. The figure in bold is a parallelogram. Find the value of *x*. Explain your thinking. (The diagram is a sketch. Do not assume that it is drawn to scale.)

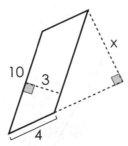

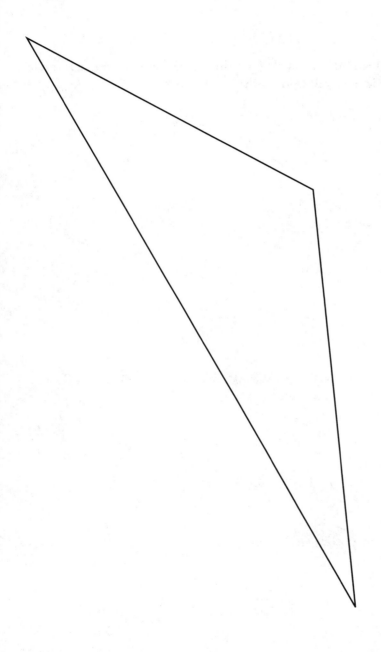

TEACHER'S GUIDE

STAGE 1

Problem #1

1. Draw a parallelogram like the one below. Show how to find a formula for its area using your knowledge of rectangle area.

Questions and Conversations for #1

» *Can you decompose the parallelogram and rearrange it?* Yes. How can you use what you already know about areas of rectangles?
Note. Use the remaining questions after students have formed a rectangle. (See the drawings in the Solution for #1.)

» *How do the sides of the parallelogram and rectangle compare?* Some of their sides are the same length, but others are not.

» *Can you use the side lengths as your variables? Why or why not?* Apply your answer to the question above.

» *Why does the term* height *make sense?* When the base is placed horizontally at the bottom of the parallelogram, the *height* represents the vertical distance from the base to the top. This fits well with the everyday meaning of the word *height*.

> **Teacher's Note.** This is an appropriate time to discuss the terms *base* and *height*. Reserve the word *base* for the side length that the parallelogram and the rectangle have in common. The *height* (of both shapes) is the length of the other side of the rectangle. Notice that you can use any segment between the bases that is perpendicular to them to find the height.

» *Was it necessary to choose the bottom side of the parallelogram as the base?* No! Think about how you would draw the rectangle if you chose a different side as the base! (See Problems #8 and #9 to explore this idea further.)

» *Once you have a formula, how can you test it?* One possibility is to draw a variety of parallelograms on a piece of graph paper. Use the grid to find or estimate their areas. Then apply your formula.

Solution for #1

The standard formula for the area of a parallelogram is $A = b \cdot h$.

Most students remove a right triangle (1) from one side and attach it to the other (2) as shown in the left drawing. The second figure suggests that you may draw a segment perpendicular to the bases at any point along the parallelogram and carry out the same process, moving region (1) over to a new location (2). In either case, you form a rectangle whose *base* has the same length as a side of the parallelogram and whose *height* is the same as the height of the parallelogram. (See Questions and Conversations for #1 above.)

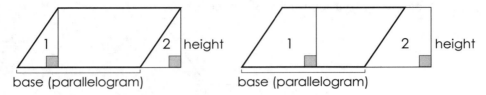

Because the area of the figure does not change when you decompose it and rearrange the parts, you may determine the area of the parallelogram by finding the area of the rectangle. Because the parallelogram and the rectangle have the same base length and the same height, a formula for the area of the parallelogram is $A = b \cdot h$. This formula applies to the rectangle as well (which is reasonable, because a rectangle is a special type of parallelogram).

Problem #2

2. Draw a triangle like the one below. Show how to find a formula for its area using your knowledge of rectangle area. Can you use the side lengths as your variables? If not, what could you use instead? Explain.

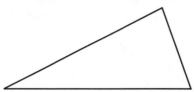

Questions and Conversations for #2

Use the Questions and Conversations for #1 with minor modifications to guide discussions about the triangle. Notice that the picture for Solution #2 shows that the *height* is also equal to the length of the *altitude* of the triangle—the perpendicular segment from the base to its opposite vertex. (Again, the terms *base* and *height* apply to both the rectangle and the triangle.) The freedom to use any side as a base is explored further in Problem #7.

Solution for #2

The area of this triangle can be found from the formula $A = \frac{1}{2} \cdot b \cdot h$.

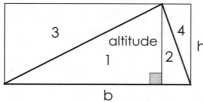

Most students surround the triangle with a rectangle as shown. The *altitude* of the triangle (see Questions and Conversations above) separates it into regions 1 and 2, which are congruent to triangles 3 and 4 respectively. Therefore, the area of the triangle is half the area of the surrounding rectangle. Because the rectangle has an area equal to its base times its height, a formula for the triangle's area is $A = \frac{1}{2} \cdot b \cdot h$.

Problem #3

3. Show how to create as many different (noncongruent) polygons as you can by joining two copies of this triangle side to side. Choose one of your drawings and explain how to use it to find a formula for the area of the triangle. Is it the same as the formula from Problem #2?

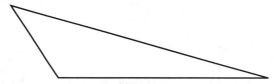

Questions and Conversations for #3

» *Can you find this triangle's area using a "surround and take half" strategy as in Problem #2?* Yes, but only if you use the longest side as the base. Can you see why? Try it if you are curious, but we will explore a different approach here.

» *How many different polygons do you think there will be? Why?* Students' answers will vary. Many will say 3 (one for each side).

» *Can you think of any geometric transformations that might help?* If students aren't sure, consider asking them to talk specifically about how they are moving the triangles as they explore. Some may suggest turns (rotations) and flips (reflections).

> **Teacher's Note.** Students may find it helpful to draw and cut out two copies of the triangle and rearrange them by hand.

» *How can you use the transformations to keep one side of the triangle fixed in position?* Reflect the triangle over the side that should not move, or rotate the triangle about the midpoint of this side.

» *Do you see any patterns in your results?* Each type of transformation results in a particular type of quadrilateral.

Teacher's Note. Once students find a formula, remember to encourage them to test it!

Solution for #3

You can form six polygons. The top three figures are parallelograms formed by rotating the triangle about the midpoint of the side with the dashed line. The bottom three figures are kites formed by reflecting the triangle over the same side.

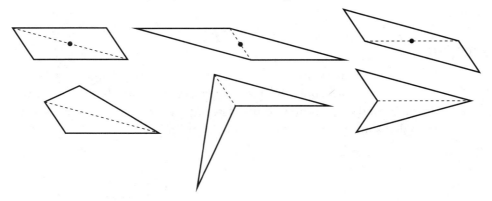

Because students found a formula for the area of a parallelogram in Problem #1, it makes sense to choose a parallelogram to work with. Within each of the top three drawings, the bases of the original triangle and the parallelogram are the same length. The same can be said of the heights. Now the area of the triangle is half of the area of the parallelogram ($A = b \cdot h$). Therefore, a formula for the area of the triangle is $A = \dfrac{1}{2} \cdot b \cdot h$. This is the same as the formula for the triangle in the previous question!

Problem #4

4. Draw a trapezoid like the one below. Apply your knowledge of previously discovered formulas to find a formula for its area, determining the variables that are needed as you work. Explain your thinking.

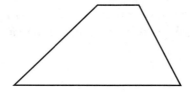

Questions and Conversations for #4

» *How many variables do you think your formula will need?* It will need three. In addition to the height, both of the parallel segments play a role in determining the area. (These are both called *bases*, and may be labeled B and b, or b_1 and b_2.)

» *How can you use the trapezoid to create a quadrilateral(s) for which you already have an area formula?* What types of strategies have you used before? (Consider decomposing and rearranging, using transformations, joining copies, or surrounding and subtracting.)

Solution for #4

The formula may be written many ways, including:

$$A = \frac{1}{2} \cdot (B+b) \cdot h \ \text{ or } \ A = \frac{1}{2} \cdot B \cdot h + \frac{1}{2} \cdot b \cdot h$$

Two common strategies are based on the following pictures. (See the Questions and Conversations for a discussion of B and b.)

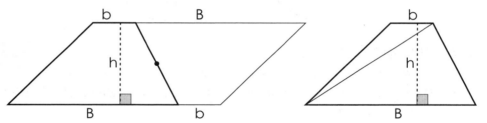

Sample strategy 1. In the first picture, the trapezoid is turned upside-down and joined to a copy of the original (or rotated about the midpoint of a side that is not a base) to create a parallelogram. The base of this parallelogram is $B+b$ and its height is h (the same as the height of the trapezoid). Therefore, the area of

> **Teacher's Note.** Students may be interested to note that the expression $\frac{1}{2} \cdot (B+b)$ is just the mean (average) of the lengths of the two bases!

the parallelogram is $A = (B+b) \cdot h$. Because the trapezoid has half the area of the parallelogram, it has an area formula of $A = \frac{1}{2} \cdot (B+b) \cdot h$.

Sample strategy 2. In the second picture, the trapezoid is decomposed into two triangles, which have base lengths B and b. Both triangles have a height of h. Thus, the areas of the triangles are given by the expressions $\frac{1}{2} \cdot B \cdot h$ and $\frac{1}{2} \cdot b \cdot h$. The area of the trapezoid is the sum of these: $A = \frac{1}{2} \cdot B \cdot h + \frac{1}{2} \cdot b \cdot h$.

Other strategies: Some students may *truncate* a triangle ("chop its top off") parallel to a base to create the trapezoid. Others may decompose the trapezoid into a rectangle and two right triangles, or surround it with a rectangle and subtract the areas of two right triangles. Students may learn something from exploring these strategies. However, in order to use them to produce a formula without additional variables, they may have to carry out further creative manipulations of their drawings or use algebraic procedures (involving the distributive property).

Problem #5

5. Use the first drawing to explain why an expression for the area of a circle appears to be between $3 \cdot r^2$ and $4 \cdot r^2$ (but closer to $3 \cdot r^2$). Use the second drawing to explain why the formula for the area of a circle is $A = \pi \cdot r^2$.

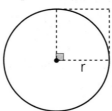 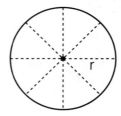

Questions and Conversations for #5

» *What is the area of the square in the first drawing?* $r \cdot r$ or r^2
» *How many of these squares would it take to cover the entire circle?* Four of them would cover it with room to spare.
» *Do you think that three of the squares would completely cover the circle?* If you don't feel sure, make a copy of the picture. Cut out some of the parts and rearrange them.
» *Is it possible to arrange the eight regions in the second picture to form a familiar shape?* Probably not, but you might be able to come close. Experiment. *Note.* Ask the next question only after students have created the first picture in the Solution below.

» *Can you think of a way to make the figure look more like a parallelogram?* Try cutting the circle into smaller pieces!

Solution for #5

The area of the square in the first picture is r^2. Four of these have an area of $4 \cdot r^2$ and would cover the entire circle with some pieces left over. Three of the squares appear not to have quite enough area, but it you cut off the regions outside the circle for these squares, and place them in the uncovered quarter of the circle, they would cover a substantial fraction (but not all) of it. Thus, it appears that the area of the circle is a little more than $3 \cdot r^2$.

The second circle above shows 8 regions (*sectors*) that can be cut apart and rearranged as shown in the first picture below. The result looks somewhat like a parallelogram!

In the picture to the right, the circle was decomposed into 16 congruent sectors and arranged the same way. It now looks even more like a parallelogram, because the top and bottom "sides" look more like line segments. In addition, the interior angles are closer to 90° than before. In fact, as you make the sectors smaller and smaller, the figure becomes virtually indistinguishable from a rectangle. Because the area of this "rectangle" is the same as the area of the circle, you may find it by multiplying its "base" and "height."

From the way that the figure was formed, you can see that its base is equal to half the circumference of the circle ($\pi \cdot r$) and that its height is equal to the radius (r). Therefore, the area of the circle is $A = (\pi \cdot r) \cdot r = \pi \cdot (r \cdot r) = \pi \cdot r^2$ Because π is slightly larger than 3, this is consistent with the prediction that the area is a little more than $3 \cdot r^2$.

STAGE 2

Problem #6

6. Develop a formula for the area of a (convex) kite in terms of the lengths of its two diagonals, d_1 and d_2. Show how you found the formula and why it works. Does your formula apply to concave kites? Explain.

Questions and Conversations for #6

»   *What does the phrase "in terms of" mean?* This is a common phrase in mathematics. In this case, it means to use the lengths of the diagonals as the variables in your formula.

»   *What do you notice about the relationship between the diagonals?* They appear to be perpendicular. (And, in fact, they are. See Exploration 2, Problem #16.)

Solution for #6

The area of the kite is given by the formula $A = \dfrac{1}{2} \cdot d_1 \cdot d_2$. Students will try variations on numerous strategies. Two of the most common strategies are shown here.

Sample solution 1. Surround the kite with a rectangle. Notice each right triangle outside the kite pairs with a congruent triangle inside the kite. Thus, the area of the kite is half that of the rectangle.

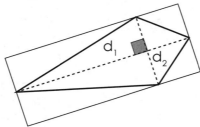

Because the rectangle has a base and height of d_1 and d_2, its area is $d_1 \cdot d_2$. Because the area of the kite is half of this, its area formula is $A = \frac{1}{2} \cdot d_1 \cdot d_2$.

Sample solution 2: The kite consists of two congruent triangles, each with a base of d_1 and a height of $\frac{1}{2} \cdot d_2$. The area of each triangle is:

$$A = \frac{1}{2} \cdot b \cdot h = \frac{1}{2} \cdot d_1 \cdot \left(\frac{1}{2} \cdot d_2 \right) = \left(\frac{1}{2} \cdot \frac{1}{2} \right) \cdot d_1 \cdot d_2 = \frac{1}{4} \cdot d_1 \cdot d_2$$

The middle steps show the associative and commutative properties of multiplication, which allow you to regroup and reorder the factors respectively. Because there are two of these triangles, the kite has an area twice this large, or:

$$A = \frac{1}{2} \cdot d_1 \cdot d_2$$

Teacher's Note. If students do not know procedures for multiplying fractions, they can think in terms of "half of d_2," "half of half," etc. When doubling $\frac{1}{4} \cdot d_1 \cdot d_2$ to obtain the final formula, they may simply ask the question "What happens if you have one-fourth of something and you double it?"

The formula for the area of a concave kite is the same, even though one of the diagonals (d_2) lies outside the figure! Many students will decompose the kite into two congruent triangles as in Sample solution 2. Imagine transforming a convex kite into a concave kite by pushing two congruent sides inward as shown by the dashed segments.

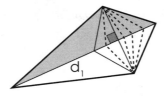

As you do this, d_1 gets smaller (as do the two sides of the kite, at first), while d_2 remains the same—but both remain diagonals of the kite. Even as the kite changes, each triangle (beginning, for example, with the one in grey) continues to have a base of d_1 and a height of $\frac{1}{2} \cdot d_2$. Therefore, the area of the kite still has the same formula!

Problem #7

7. Find the area of the triangle on p. 80 as accurately as you can. Carry out the process three times, using a different side as the base each time. Make any necessary additions to the drawing as you work. Report your answers with an appropriate level of precision. Do all three answers agree?

Questions and Conversations for #7

»   *What units should you use for your measurements?* Centimeters, millimeters, or inches are appropriate for the size of the triangle. Which will give you more precision? Which will be easier to round? Do you know how to multiply mixed numbers? (Given the answers to these questions, most students will probably choose centimeters or millimeters.) If some choose inches, ask students to compare the advantages and disadvantages of each after they've finished calculating.

»   *What are some techniques for determining the heights? Which are likely to be the most accurate?* Some students draw a line parallel to the base through the opposite vertex and then measure the (perpendicular) distance between this line and the base. Others extend a base and draw a perpendicular segment from this extension to the opposite vertex. The latter method is shown in the solution. It is usually more accurate.

»   *How close should your three answers be?* If you draw and measure carefully, the answers should agree within approximately 1 cm^2.

»   *How precisely should you report your answer?* This depends on your answer to the previous question. For example, if your answers are in approximate agreement to the nearest cm^2, then you should round to the nearest cm^2.

»   *What can you say about the height that corresponds with the largest base? Why?* The largest base must have the smallest height because the product of the base and height is the same (twice the area of the triangle) for every base/height pair.

Solution for #7

The sketch below shows base and height measurements (cm) and altitudes (dashed segments). Notice how two of the bases were extended to help draw altitudes. Students must be careful to match each base with its corresponding height.

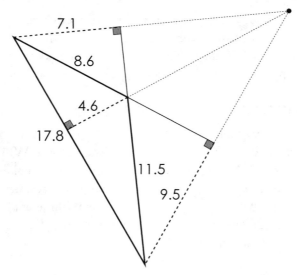

Teacher's Note. If you extend the three altitudes (dotted segments), they should meet at a single point (known as the orthocenter of the triangle). Some students may be interested in testing the accuracy of their drawing by trying this!

The calculations below are likely more precise than most students will manage.

$$A = \frac{1}{2} \cdot b \cdot h = \frac{1}{2}(17.8)(4.6) \approx 40.9 \text{ cm}^2$$

$$A = \frac{1}{2} \cdot b \cdot h = \frac{1}{2}(11.5)(7.1) \approx 40.8 \text{ cm}^2$$

$$A = \frac{1}{2} \cdot b \cdot h = \frac{1}{2}(8.6)(9.5) \approx 40.9 \text{ cm}^2$$

Students should use the variation in their three answers to help them determine how precisely to report their results. The results above suggest reporting a result to the nearest tenth of a square centimeter (40.9 cm²). However, students' measurements may not be this precise, in which case they should round to the nearest whole number. The best answer is probably 41 cm². Given the challenge in measuring precisely and drawing altitudes accurately, answers between about 39 cm² and 42 cm² are probably reasonable.

STAGE 3

Problem #8

8. Show how to find the exact area of this parallelogram (in units^2) without using a formula. Would the formula from Problem #1 still work? Why or why not?

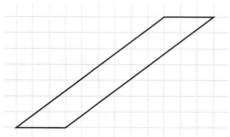

Questions and Conversations for #8

» *What makes this more challenging than the parallelogram in Problem #1?* The parallelogram is very "tilted!" You cannot make a cut that is perpendicular to the (horizontal) base and that extends all the way from one base to the other within the parallelogram.

» *Are there ways to decompose and rearrange the parallelogram to make it look more like the one in Problem #1?* Yes, there are many ways to do this.

» *Is it possible to make use of multiple copies of the parallelogram?* Yes. How can you arrange them in order to get a figure that is easier to work with? How can you compensate for the fact that the area may have changed?

For students who attempt Strategy 1 below:

» *How can you rotate the parallelogram accurately?* Consider using a compass, protractor, and/or ruler to carry out the rotation. Do not assume that all vertices and sides will still align with the grid!

For students who have already decomposed and rearranged the parallelogram to create a new figure:

» *How do the base and height of your new figure compare with that of the original parallelogram?* Answers may vary, but in many cases, they will be the same!

Solution for #8

Sample strategy 1. Rotate the parallelogram so that a long side is horizontal in order to make it easier to visualize this side as a base. Slide a triangle to the opposite side of the parallelogram (as in Problem #1) to create a rectangle. The rectangle appears to have a base of about 11.3 units and a height of about 1.9 units. These measurements give an estimated area of $A \approx (11.3)(1.9) \approx 21.5$ units^2. This is close to the value of 21 units² that you would get by choosing the smaller side as the base (3 units) and multiplying this by the corresponding height (7 units). This suggests that the area formula from Problem #1 may still apply.

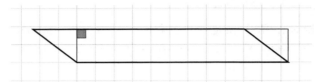

This method demonstrates students' understanding that any side of the parallelogram may serve as a base. However, it does not provide an exact area as stipulated in the problem, because some vertices of the parallelogram do not lie on grid lines! (See Exploration 6 to explore this idea further.) The method also fails to help you understand why you would obtain a correct value for the area by using a short side as the base.

Sample strategy 2. (1) Slice the parallelogram horizontally along the grid lines into smaller parallelograms. (2) Align the small parallelograms atop one another. (3) Cut and slide a right triangle from each small parallelogram to form a rectangle. Because the areas of the parallelogram and the rectangle are the same, the area of the parallelogram is $3 \cdot 7 = 21$ units². Because the base and height of the rectangle are equal to the base and height of the parallelogram, the formula for the parallelogram's area appears to be $A = b \cdot h$, as before.

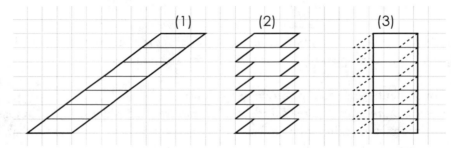

Sample strategy 3. (1) Join multiple copies of the parallelogram side by side until you get a parallelogram for which a cut perpendicular to the base extends all the way from one base to the other within the parallelogram. (2) Make this type of cut. (3) Slide the cut (grey) piece to the right to from a rectangle.

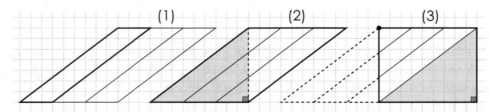

The height is unchanged in this process. However, because three copies of the parallelogram were used to create the rectangle, its base is now three times as long. Thus, the area of the parallelogram is one-third the area of the rectangle. Because the area of the rectangle is $9 \cdot 7 = 63$ units^2, the area of the parallelogram is $\frac{1}{3} \cdot 63 = 21$ units^2. This shows that the area formula from Problem #1 still works, because the area of the rectangle is $3 \cdot b \cdot h$, and the area of the parallelogram is one-third of this, or $b \cdot h$. The same argument applies no matter how many parallelograms you join.

Note that this problem leaves a lot of scope for students' creativity! For example, some may slice the original parallelogram horizontally in the center and join the pieces to try to create a parallelogram that is less "tilted." Others may experiment with "stair-step" patterns inside the parallelogram.

Problem #9

9. The figure in bold is a parallelogram. Find the value of *x*. Explain your thinking. (The diagram is a sketch. Do not assume that it is drawn to scale.)

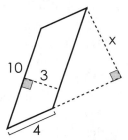

Questions and Conversations for #9

» *How can you apply what you learned in the previous question?* Try looking at the parallelogram from different perspectives! (Some students may find it helpful to physically turn the page.)

Teacher's Note. Be sure to give students plenty of time to think about this problem before giving guidance. Although the solution process is short, many students will need time to realize that you can apply the idea of using different sides as the base. When they begin, they tend to spend time focusing on the triangle at the right.

Solution for #9

You can calculate the area of the parallelogram using its long side as the base: $A = 10 \cdot 3 = 30$ units^2. If you use the short side as the base instead, the corresponding height is *x*. Consequently, the area is also given by $4 \cdot x$. Equating the two expressions for the area: $4 \cdot x = 30$. The value of *x* is $30 \div 4 = 7.5$ units.

WRAP UP

Share Strategies

Give students an opportunity to share formulas and strategies for finding them. If the formulas differ, use the opportunity to discuss the concept of writing algebraic expressions in different, but equivalent forms. If the strategies differ, compare them. Do they make sense? Are some strategies more efficient? Easier to understand? Do they give you different insights into the mathematics?

Summarize

Answer any remaining questions that students have. You may also want to summarize a few key ideas:

» Beginning with the formula $A = b \cdot h$ for rectangles, you can develop formulas for other geometric figures using strategies such as decomposing and rearranging, surrounding and subtracting, joining copies of the figures, etc. Each time you create a new formula, you can use it to create others.

» *Base* and *height* are measurements that are closely related to area for a variety of geometric figures. You often have the option of choosing a side to use as a base. The base and height are always measured at right angles to each other.

» There are always multiple equivalent ways to write expressions and formulas.

» When you report the results of calculations based on measurements, you must take into account the precision of the measurements.

Further Exploration

Ask students to think of new questions to ask or ways to extend this exploration. Here are some possibilities:

» Find a formula for the area of a circle using the diameter or the circumference as a variable. (*Answers:* $A = \frac{1}{4} \cdot \pi \cdot d^2$; $A = \frac{c^2}{4 \cdot \pi}$)

» A *midsegment* of a triangle is a line segment joining the midpoints of two sides of the triangle. Extend this definition so that it applies to parallelograms and trapezoids as well. Develop area formulas for each shape using only the midsegment (*M*) and height (*h*) as variables. Do any of your formulas apply to squares, rectangles, or kites? Why or why not? (*Partial answer:* Parallelograms, triangles, and trapezoids have the same simple formula! $A = M \cdot h$)

» Suppose that equilateral triangles are used as the basic unit for measuring area. How does this affect the formulas for familiar geometric figures? Which shapes have more complicated formulas? Which become simpler?

Which (if any) parts of the formulas stay the same? Which parts change? Are there other shapes you could use as a basic unit? Are there shapes you could never use as a basic unit? (*Note.* If you return to this question after completing Exploration 6, you may be able to make further progress!)

» Suppose a triangle in a coordinate grid has vertices at (x_1, y_1), (x_2, y_2), and (x_3, y_3). Find a formula for the area of the triangle using these coordinates as variables. (It is likely to be a complex formula, but if you know some algebra, including the distributive property, you can simplify it quite a bit!) Does your formula depend on which ordered pair is assigned to each vertex?

Exploration **6**

A New Slant on Measurement

INTRODUCTION

Materials

- » Graph paper
- » Scientific calculator
- » Compass and straightedge to draw pictures for some problems (optional)

Prior Knowledge

- » Complete Exploration 4: Geoboard Squares, Stage 1.
- » Complete Exploration 5: Area Formulas, Stages 1 and 2.
- » Estimate square roots, and evaluate them with and without a calculator.

> **Teacher's Note.** This exploration is designed for use before students have been exposed to the Pythagorean Theorem.

Learning Goals

- » Use knowledge of area to develop a formula for the relationship between the sides of a right triangle.
- » Know that there are numbers (*irrational numbers*) whose decimal continues forever without settling into a permanent repeating pattern.
- » Use logic (*deductive reasoning*) to explain why the Pythagorean Theorem is true.
- » Apply the Pythagorean Theorem to solve challenging problems.
- » Pay attention to precision of numbers in mathematical and real-world problems.
- » Communicate complex mathematical ideas clearly.
- » Persist in solving challenging problems.

Launching the Exploration

Motivation and purpose. To students: You may have noticed that the lengths of "slanted" segments on a grid are not always whole number multiples of the grid's basic unit. In this exploration, you discover and apply a procedure to help you find these lengths. The ideas you learn here will support your future understanding of central concepts in trigonometry, geometry, and calculus!

Understanding the problem. Review "tilt numbers" and strategies for finding areas on graph paper. (See Exploration 4: Geoboard Squares.) Skim through the Student Handout to give students a sense for what it entails, but do not give away the fact that it is about the Pythagorean Theorem!

Teacher's Note. This activity does not provide routine exercises for students to practice using the formula $a^2 + b^2 = c^2$. Some students may benefit from such practice before attempting more challenging problems such as those in Stages 2 and 3.

Feel free to introduce the vocabulary terms *Pythagorean Theorem, leg,* and *hypotenuse* at any point after Problem #5. (See the Summarize portion of the Wrap Up.) This terminology does not appear in the Teacher's Guide until Stage 3.

STUDENT HANDOUT

Stage 1

1. Calculate or estimate the side lengths of squares having areas of 16, 289, and 11 square units. Explain your thinking.

2. On graph paper, draw squares having sides with the given "tilt numbers" (see Exploration 4). Show strategies for calculating their areas. Use the area to find the length of each side of the square. You may give approximate values for some answers.

   a. 1, 1   b. 1, 2   c. 3, 4   d. 2, 5   e. 6, 1   f. 7, 5

3. Make a table of your results in Problem #2. Use it to find a formula for the area of a square in terms of the tilt numbers, a and b, of its sides.

4. Starting at home, you ride your bike to a friend's house along the streets in your neighborhood. You travel approximately 73 meters to the east and then 180 meters to the north. By about how much would it reduce the total length of your trip if there were a street that took you *directly* to your friend's house along a straight path? Make a drawing of the situation, and use it to explain your thinking process.

5. Sketch a right triangle and label the lengths of its two shorter sides a and b. Label the longest side length as c. Write a formula that describes the relationship between a, b, and c. Explain your thinking process. What happens if the triangle does not have a right angle?

Stage 2

6. The pictures below show two ways to decompose a square. Use the pictures to explain why the equation you found in Problem #5 must always be true for right triangles.

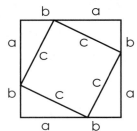 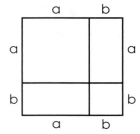

7. Find the area of an equilateral triangle whose sides have lengths of 8 feet.

8. The picture below shows a square *inscribed* in a circle. If the side lengths of the square are 6 cm, calculate the approximate area of the shaded region.

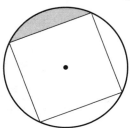

Stage 3

9. The following picture shows two overlapping circles that intersect at the points *A* and *B*. One circle is centered at *O* and the other at *P*. The smaller circle has a radius of 1 unit. Calculate the area of the shaded *lune*. (Can you guess why it is called a *lune*?)

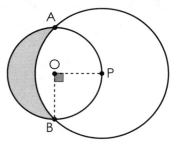

10. You are flying a kite and have let its 30-meter string out to its full length. The kite is above a point on the ground that lies about 13 meters north and 9 meters east of the place you are standing. Make a sketch of this situation, and use it to determine the approximate height of the kite.

TEACHER'S GUIDE

STAGE 1

Problem #1

1. Calculate or estimate the side lengths of squares having areas of 16, 289, and 11 square units. Explain your thinking.

Questions and Conversations for #1

This section contains ideas for conversations, mainly in the form of questions that students may ask or that you may pose to them. Be sure to allow students to do most of the thinking and talking!

» *Should you use a calculator?* Depending on the value of the area, you may not need a calculator. If you think you do, try to estimate the side length before using it.

Solution for #1

The side lengths of the squares are 4, 17, and approximately 3.32 units. Because the area of a square is equal to its side length multiplied by itself, you can calculate each side length by finding the square root of the corresponding area.

$$\sqrt{16} = 4 \qquad\qquad \sqrt{289} = 17 \qquad\qquad \sqrt{11} \approx 3.32$$

Problem #2

2. On graph paper, draw squares having sides with the given "tilt numbers" (see Exploration 4). Show strategies for calculating their areas. Use the area to find the length of each side of the square. You may give approximate values for some answers.

a. 1, 1   b. 1, 2   c. 3, 4   d. 2, 5   e. 6, 1   f. 7, 5

Questions and Conversations for #2

» *Does the order of the tilt numbers matter? Why or why not?* No. You will get the same (congruent) square regardless of the order of the tilt numbers.

» *If an answer has many decimal places, to what precision should you show it?* If you regard the tilt numbers as having been measured, then the level of precision (number of decimal places) with which you measured them should be approximately the same as the precision you report in your solution. If you think of the tilt numbers as being exact, you could show as many decimal places as you like. In the solutions, I show two places.

» *How many decimal places would you need in order to get an exact answer?* The square root of a nonsquare whole number is an *irrational* number, which has an infinite number of decimal places. Your answer will not be exact no matter how many decimal places you write.

Solution for #2

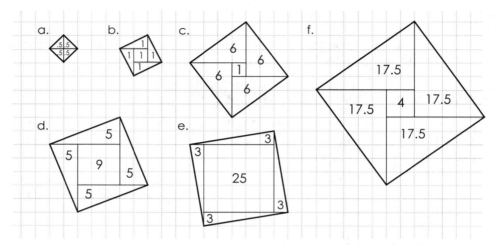

These diagrams show a decomposition strategy. Some students may use a surround and subtract or other strategy.

a. Area: $0.5 + 0.5 + 0.5 + 0.5 = 2$ units^2 Length of Side: $\sqrt{2} \approx 1.41$ units
b. Area: $1 + 1 + 1 + 1 + 1 = 5$ units^2 Length of Side: $\sqrt{5} \approx 2.24$ units
c. Area: $1 + 6 + 6 + 6 + 6 = 25$ units^2 Length of Side: $\sqrt{25} = 5$ units
d. Area: $9 + 5 + 5 + 5 + 5 = 29$ units^2 Length of Side: $\sqrt{29} \approx 5.39$ units
e. Area: $25 + 3 + 3 + 3 + 3 = 37$ units^2 Length of Side: $\sqrt{37} \approx 6.08$ units
f. Area: $4 + 17.5 + 17.5 + 17.5 + 17.5 = 74$ units^2
Length of Side: $\sqrt{74} \approx 8.60$ units

Problem #3

3. Make a table of your results in Problem #2. Use it to find a formula for the area of a square in terms of the tilt numbers, *a* and *b*, of its sides.

Questions and Conversations for #3

» *What data should you include in your table?* The data from the previous question are the tilt numbers, the area, and the side length. The tilt numbers and area are the most important, because they relate to the formula for which you are searching.

» *What types of operations are likely to appear in your formula? Why?* Students may suspect that the formula will involve squaring, because the situation involves geometric squares. Some may think that multiplication will appear, because the areas get significantly larger even when the tilt numbers do not change a great deal. Some may expect to see addition or subtraction, because the decomposition and surround/subtract strategies involve adding or subtracting areas.

» *How do you proceed if you can't find patterns in your data?* Consider gathering additional data and organizing it. For example, arrange pairs of tilt numbers as shown on the left. The areas of the corresponding squares are shown on the right.

Tilt Numbers					Areas				
1, 1	1, 2	1, 3	1, 4	1, 5	2	5	10	17	26
2, 1	2, 2	2, 3	2, 4	2, 5	5	8	13	20	29
3, 1	3, 2	3, 3	3, 4	3, 5	10	13	18	25	34
4, 1	4, 2	4, 3	4, 4	4, 5	17	20	25	32	41
5, 1	5, 2	5, 3	5, 4	5, 5	26	29	34	41	50

Notice that the areas in the first row (and column) are all one greater than a square number. What about the second row? There are many interesting and useful patterns here, some of which can lead to a formula!

» *Can you apply general strategies for finding areas to help you obtain a formula?* Yes! Decomposition or surround and subtract strategies may help, although they may not produce the simplest version of the formula.

» *What does a square look like if one of its tilt numbers is 0? Does your formula still apply in this case?* Its sides will be vertical and horizontal. Try it! Your formula should still work!

Solution for #3

a	b	Area (units²)	Side Length (units)
1	1	1	$\sqrt{1} = 1$
1	2	5	$\sqrt{5} \approx 2.24$
3	4	25	$\sqrt{25} = 5$
2	5	29	$\sqrt{29} \approx 5.39$
6	1	37	$\sqrt{37} \approx 6.08$
7	5	74	$\sqrt{74} \approx 8.60$

Students who search for the formula by analyzing patterns in the table are likely to discover the result $A = a^2 + b^2$. (Notice that the capital A and the lowercase a are different variables. A represents area, while a stands for one of the tilt numbers.) Those who use their diagrams from Problem #2 may find other formulas such as

$$A = (a-b)^2 + 2 \cdot a \cdot b \text{ (from the decomposition strategy), or}$$

$$A = (a+b)^2 - 2 \cdot a \cdot b \text{ (from the surround and subtract strategy)}$$

In the first of these formulas, $(a-b)^2$ represents the area of the inner square, while $2 \cdot a \cdot b$ stands for the area of the 4 surrounding triangles. (See solutions to Problem #2.) In the second formula, $(a+b)^2$ is the area of the surrounding square, and $2 \cdot a \cdot b$ represents the total area of the four triangles whose area must be subtracted from this.

Problem #4

4. Starting at home, you ride your bike to your friend's house along the streets in your neighborhood. You travel approximately 73 meters to the east and then 180 meters to the north. By about how much would it reduce the total length of your trip if there were a street that took you *directly* to your friend's house along a straight path? Make a drawing of the situation, and use it to explain your thinking process.

Teacher's Note. Students are asked to solve Problem #4 before developing a complete formula in Problem #5. This challenges them to develop a deeper and more flexible understanding of the concepts in Problems #2 and #3 and to make a connection to right triangles before formalizing the concept algebraically.

Questions and Conversations for #4

» *What connections do you see between your picture of this situation and your drawings in Problem #2?* The direct route corresponds to the "tilted" segment that is used to make the square. The paths along the streets correspond to the segments that show the tilt numbers.

» *What could you add to your picture of this situation in order to help you visualize the connections to Problem #2?* You may find it helpful to draw a square using the direct path as one of the sides.

Solution for #4

The distance would be reduced by approximately 59 meters.

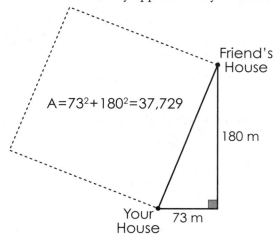

Think of the direct path from your house to your friend's house as the side of a square whose tilt numbers are 73 and 180. Then, based on the results of Problem #3, the area of the square is $73^2 + 180^2 = 5329 + 32,400 = 37,729$ square meters. The direct distance to your friend's house is the length of one side of this square:

$$\sqrt{37,729} \approx 194 \text{ meters}$$

> **Teacher's Note.** Some students will be able to carry out this reasoning process without showing the tilted square in their diagram.

Following the streets, you would travel $73 + 180 = 253$ meters. The distance would be reduced by approximately $253 - 194 = 59$ meters if you were able to take the direct route.

Problem #5

5. Sketch a right triangle, and label the lengths of its two shorter sides *a* and *b*. Label the length of the longest side as *c*. Write a formula that describes the relationship between *a*, *b*, and *c*. Explain your thinking process. What happens if the triangle does not have a right angle?

Questions and Conversations for #5

»   *How can you use Problem #4 as a model to create a picture for this problem?* Replace numbers with variables.

»   *How can you write the area of the tilted square as an algebraic expression?* Use the variable *c*, and apply what you know about areas of squares.

» *What is the relationship between the area of the tilted square and the tilt numbers?* This is the relationship you discovered in Problem #3. Your job is to express it algebraically.

» *What happens to the length of the third side when the angle between the two shorter sides of the triangle changes?* It changes as well. How does the change in the angle relate to the change in the side length?

Solution for #5

One way to write the formula is $c^2 = a^2 + b^2$.

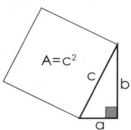

The drawing above is like the one in the previous question, but with the numbers replaced by the variables a, b, and c. The area of the tilted square (c^2) is the sum of the squares of the tilt numbers, a and b ($a^2 + b^2$). Some students may create a formula that gives the actual value of c by incorporating the process of taking the square root: $c = \sqrt{a^2 + b^2}$.

If the triangle does not have a right angle, then a and b will no longer be tilt numbers, and you cannot expect the formula to remain true. In fact, if the angle between the two shorter sides is acute, then the third side will decrease in length, making c^2 less than $a^2 + b^2$. By the same type of reasoning, if this angle is obtuse, c^2 will be greater than $a^2 + b^2$.

Teacher's Note. Consider asking students what happens if you draw squares from the other two sides of the triangle as well. The area of the largest square is the sum of the areas of the two smaller squares!

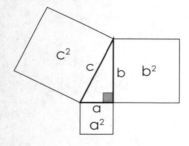

STAGE 2

Problem #6

6. The pictures on the next page show two ways to decompose a square. Use the pictures to explain why the equation from Problem #5 must always be true for right triangles.

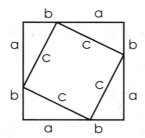 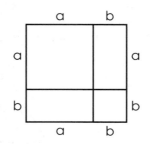

Questions and Conversations for #6

» *What happens if you remove equal areas from both figures?* Because the original squares have equal areas, the areas that remain will also be equal.

Teacher's Note. Some students may benefit from making an accurate copy of these drawings, cutting the squares into pieces, and rearranging them.

» *How do you know that the tilted figure inside the square on the left is a square?* The sides are all the same length (c), but how do you know that the interior angles are all congruent? (What can you say about the sum of the two acute angles in each triangle?)

Solution for #6

The two squares have the same area. Thus, if you remove the same area from each, the remaining areas must be equal. The four right triangles on the left have the same total area as the two (nonsquare) rectangles on the right. (Can you see why?) If you remove these, it shows that the area of the tilted square on the left (c^2) and the sum of the areas of the two squares on the right ($a^2 + b^2$) must be equal. Therefore, $c^2 = a^2 + b^2$!

To justify this more completely, you need to verify that the tilted figure on the left is actually a square. Its sides are all the same length, but are the interior angles all 90°? Notice that the two acute angles in each right triangle must be *complementary* (they have a sum of 90°). Choose one of the points on the large square where two triangles meet. There are three angles that combine to make a straight angle (180°). Two of them are the complementary angles described above. Therefore the remaining angle (in the tilted square) must have a measure of $180° - 90° = 90°$.

Problem #7

7. Find the area of an equilateral triangle whose sides have lengths of 8 feet.

Questions and Conversations for #7

» *What might you add to your drawing of the triangle?* Consider drawing an altitude to help you find the height. How does this enable you to apply what you have learned earlier in this activity?

Solution for #7

The area of the equilateral triangle is $4 \cdot \sqrt{48}$, or approximately 27.7 ft².

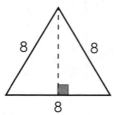

When you draw the altitude of the triangle, it forms two (congruent) right triangles, because the altitude is perpendicular to the base. One of these is shown below.

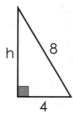

The height of the equilateral triangle is the side length of this right triangle labeled h. According to the formula from Problem #5, the relationship between the sides is:

$$4^2 + h^2 = 8^2 \text{ or } 16 + h^2 = 64$$

Teacher's Note. If students use a different strategy to find the area, or if they know that $\sqrt{48}$ is equal to $4 \cdot \sqrt{3}$ (see Exploration 10: Factor Blocks and Radicals in the *Numbers and Operations* book from this series), they may write the area in the equivalent form $16 \cdot \sqrt{3}$.

This means that $h^2 = 48$. Therefore, the height of the equilateral triangle is $\sqrt{48}$ (approximately 6.9) feet.

To find the area of the equilateral triangle, take half the product of its base and height:

$$A = \frac{1}{2} \cdot 8 \cdot \sqrt{48} \text{ or } A = 4 \cdot \sqrt{48}.$$

This is approximately equal to 27.7 ft².

Problem #8

8. The picture below shows a square *inscribed* in a circle. If the side lengths of the square are 6 cm, calculate the approximate area of the shaded region.

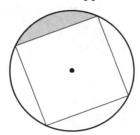

Questions and Conversations for #8

»   *What does* inscribed *mean?* In this case, it means that the square is drawn inside the circle in such a way that each of its vertices lies on the circle.

»   *Can you estimate the area first?* By recognizing that the square has an area of 36 units², you will be able to make a rough estimate. For example, if you believe that the area of the shaded region appears to be approximately one-sixth that of the square, you might estimate a value of about 6 units².

»   *What might you add to the drawing to show connections between the square and the circle?* Because you want to calculate the circle's area, consider drawing its diameter or radius. How can you do this in a way that shows a connection to the square?

Solution for #8

The most common strategy is to subtract the area of the square (36) from the area of the circle ($18 \cdot \pi$), and divide the result by 4. This results in the expression $\dfrac{18 \cdot \pi - 36}{4}$ or approximately 5.1 units². (This is reasonably close to the estimate from the Questions and Conversations section above.)

The most challenging part of the problem is to find the area of the circle.

Sample strategy 1. The diameter of the circle is the length of a diagonal of the square. When you draw this diagonal, it forms two (congruent) right triangles that look like this, where *d* represents the diameter.

The formula from Problem #5 gives the relationship $d^2 = 6^2 + 6^2 = 72$. Therefore, the diameter is $\sqrt{72}$. The radius is half of this: $\dfrac{1}{2} \cdot \sqrt{72}$. (Most students will probably not recognize that this expression is equivalent to $\sqrt{18}$, but if they enter it into their calculator and square the result, they will see that r^2 is 18!

Sample strategy 2. Draw a right triangle to find the radius of the circle directly.

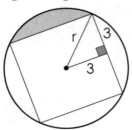

You can determine the radius from the relationship $r^2 = 3^2 + 3^2 = 18$. This means that $r = \sqrt{18}$.

Regardless of the strategy you employ, because the formula for the area of a circle contains the quantity r^2, you can directly use the fact that $r^2 = 18$!

$$A = \pi \cdot r^2 = \pi \cdot 18$$

STAGE 3

Problem #9

9. The following picture shows two overlapping circles that intersect at the points *A* and *B*. One circle is centered at *O* and the other at *P*. The smaller circle has a radius of 1 unit. Calculate the area of the shaded *lune*. (Can you guess why it is called a *lune*?)

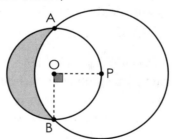

Questions and Conversations for #9

» *What would you estimate the area of the lune to be?* The area of the small circle is π because its radius is 1. The lune appears to have about one third the area of the circle. One third of π is a little greater than 1. Some students may simply estimate 1 unit².

Solution for #9

The area of the lune is exactly 1 unit^2! One way to find it is to subtract the area of the entire shaded region below (both light and dark) from the area of the small circle. The area of the shaded region is the sum of the areas of the light and dark regions.

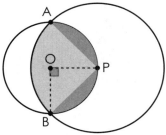

Putting this together:
Area of lune = area of small circle − area of light region − area of dark region

Area of small circle: $\pi \cdot 1^2 = \pi$

Area of light region: This is one-fourth the area of the large circle. (Can you see why?) The radius of the large circle is the length of $\overline{PB}$, which, is $\sqrt{2}$. (Use the Pythagorean Theorem with triangle $\triangle POB$.) Therefore the area of the light region is:

$$\frac{1}{4} \cdot \pi \cdot \left(\sqrt{2}\right)^2 = \frac{1}{4} \cdot \pi \cdot 2 = \frac{1}{2} \cdot \pi \text{ or } \frac{\pi}{2}$$

Area of dark region: This is the half the area of the small circle minus the area of $\triangle APB$.

$$\frac{1}{2} \cdot \pi - \frac{1}{2} \cdot 2 \cdot 1 = \frac{1}{2} \cdot \pi - 1 \text{ or } \frac{\pi}{2} - 1$$

Therefore the area of the lune is:

$$\pi - \frac{\pi}{2} - \left(\frac{\pi}{2} - 1\right) = \frac{\pi}{2} - \left(\frac{\pi}{2} - 1\right) = 1 \text{ unit}^2.$$

To get the first $\frac{\pi}{2}$ in the middle expression, subtract half of π from π. To get the final answer of 1, realize that whenever you subtract 1 less than a number from that number, the answer is always 1!

Problem #10

10. You are flying a kite and have let its 30-meter string out to its full length. The kite is above a point on the ground that lies about 13 meters north and 9 meters east of the place you are standing. Make a sketch of this situation, and use it to determine the approximate height of the kite.

Teacher's Note. If students are having trouble visualizing and drawing a picture, suggest that they imagine that the situation takes place inside a rectangular prism. (See the solution.)

Questions and Conversations for #10

» *What is a key difference between this problem and previous ones?* It takes place in three dimensions.

» *What factors might affect the precision of your answer?* You may choose not to take account of your height, and you do not know the size of the kite. Do these matter?

Solution for #10

The height of the kite is approximately 25 or 26 meters.

In this diagram, *Y* represents you, *K* stands for the kite, and *G* is the point on the ground directly below the kite. The variable *h* represents the height of the kite above the ground.

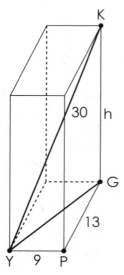

Because $\triangle YGK$ is a right triangle, you may be able to use it to find the value of *h*. Unfortunately, you don't know the length of $\overline{YG}$. However, you can find it by looking at $\triangle YPG$, which is also a right triangle! The square of the length of $\overline{YG}$ is $13^2 + 9^2 = 250$.

($\overline{YG}$ itself has a length of $\sqrt{250}$, but you will use the square of this length with $\triangle YGK$.)

Applying the Pythagorean Theorem to $\triangle YGK$ gives the relationship $250 + h^2 = 30^2 = 900$ for *h*. (Why is the 250 not squared in this formula?) Therefore, $h^2 = 650$, resulting in a height of $h = \sqrt{650} \approx 25.5$, or about 25 to 26 meters.

WRAP UP

Share Strategies

Give students an opportunity to share strategies for finding the formula in Problem #3. If they produced different formulas, ask them to compare them. Are they all correct? How can you tell? If students completed problems in Stages 2 and 3, have them compare strategies and results.

Summarize

Answer any remaining questions that students have. Summarize and expand upon some key ideas:

» If you have not already done so, name the *Pythagorean Theorem*: If a and b are the lengths of the *legs* (the two shorter sides) of a right triangle, and if c is the length of the *hypotenuse* (the longest side) of the same right triangle, then $a^2 + b^2 = c^2$. Note that the formula is not the theorem. The Pythagorean Theorem is an if/then statement that *contains* the formula.

» The *Pythagorean Theorem* is named after the mathematician Pythagoras. (You might want to do some research about him!) A *theorem* is not a theory. It is a mathematical statement that has been proved to be true.

» Some numbers (*irrational* numbers) have decimals that go on forever without settling into a permanent repeating pattern. The square roots of nonsquare whole numbers are always irrational numbers.

» Notice that the proof of the Pythagorean Theorem explored in Problem #6 is strongly connected to the surround and subtract strategy for finding areas!

» Think about how Problems #3 and #6 show that $(a+b)^2 \neq a^2 + b^2$. If a and b are positive numbers, which expression is always greater? Why?

» In Problems #2 and #3, you used patterns to *predict* a formula (*inductive* reasoning). In Problem #6, you used logic to *explain why* the formula must always be true under certain conditions (*deductive* reasoning).

» It is important to pay attention to the precision of numbers when you solve a problem. If the problem is purely mathematical (see Problems #7, #8, and #9), you often assume that the numbers are exact. In this case, you typically show as many decimal places as you feel is appropriate. If it is a real-world problem (see Problems #4 and #10) involving measured quantities, your answer should generally be rounded to the lowest level of precision of any of the measurements (to the nearest whole number for these problems).

» When you solve multistep problems that involve numbers with many decimal places, it is generally best to leave the numbers in exact form (containing the symbols for π or the square root) for as long as you can. As much as possible, avoid writing and calculating with numbers in decimal form. This prevents errors due to rounding.

Further Exploration

Ask students to think of new questions to ask or ways to extend this exploration. Here are some possibilities:

» The following picture is a common floor-tiling pattern. It illustrates a way to *tessellate* a plane—that is, to cover a plane with geometric figures in a repeating pattern that can be continued forever with no gaps or overlaps. Use this picture to give a second explanation of why your equation from Problem #5 must be true. (The four dots are a hint!)

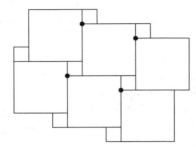

» There are many other ways to prove the Pythagorean Theorem! Do some research to learn about some of them. Even better, try to create one of your own!

» Find formulas for the areas of equilateral triangles, regular hexagons, and regular octagons based on the length of a side. Are you able to find formulas for other types of regular polygons? If so, try it. If not, explain why.

» Extend Problems #8, #9, and #10 by replacing the numbers with variables and finding a formula for each situation.

» Is it possible to create an equilateral triangle on a *geoboard*? (See Exploration 4: Geoboard Squares.) If so, do it and justify your result. If not, explain why you believe it is impossible.

» Do some research about *Pythagorean triples*.

» Use right triangles to explore equivalent radical (square root) expressions. (For example, the Solutions to Problem #8 show that $\frac{1}{2} \cdot \sqrt{72} = \sqrt{18}$. Use a picture to show why this must be true.)

Exploration 7

Ladders and Saws

INTRODUCTION

Materials

- » Ruler and protractor
- » Dynamic geometry software such as GeoGebra (optional)

Prior Knowledge

- » Know definitions of *parallelograms*, *rectangles*, *squares*, *trapezoids*, and *kites* (including the standard definition of a parallelogram as a quadrilateral in which opposite sides are parallel).
- » Know how to write names of segments, angles, and polygons.
- » Know the following vocabulary, concepts, and symbols: *congruent* ($\cong$), *parallel* ($\parallel$), *vertical* angles, *supplementary* angles, *midpoint* of a line segment, *diagonal* of a polygon.

> **Teacher's Note.** Dynamic geometry software enables students to create geometric drawings, manipulate them "on the fly," and observe the effects on measurements. It is a powerful tool that students may use in lieu of creating multiple drawings and measuring by hand. It could be used effectively on most of the problems in this exploration.

Learning Goals

- » Recognize and analyze figures embedded in complex drawings.
- » Solve problems and make/test conjectures about angle relationships involving parallel lines.
- » Apply knowledge of relationships between categories of quadrilaterals.
- » Begin exploring the roles of assumptions (*axioms* or *hypotheses*) and *converses* in mathematical reasoning.
- » Communicate complex mathematical ideas clearly.
- » Persist in solving challenging problems.

Launching the Exploration

Motivation and purpose. To students: Some geometrical drawings are quite complex. To analyze them, mathematicians often look for simpler features or patterns within them. In this activity, you will explore two particularly important patterns of this type. As you work, notice how you are using logic to build sophisticated explanations on the foundation of just a few basic observations!

Teacher's Note. This exploration may take longer than others to introduce. I recommend setting aside at least 30 minutes to discuss ladders and saws.

Understanding the problem. Discuss the distinction between the notations for geometric figures and their measurements. For example, $\overline{AB}$ represents a segment with endpoints A and B, while AB stands for the length of this segment. Similarly, $\angle CAB$ represents an angle while $m\angle CAB$ stands for the measure of this angle. Geometric figures may be *congruent*, while their measurements may be *equal*.

To explore relationships involving parallel lines, students will use the notions of *ladders* and *saws* as described in Fuys, Geddes, and Tischler (1988). Begin by showing students some examples and nonexamples of *ladders*.

Examples of Ladders:

Nonexamples of ladders:

Ask questions such as these:
»   *What similarities do you see between the ladders in the pictures and real-world ladders?* Each ladder in the drawings contains a rail and some rungs. The rungs are all parallel.
»   *What differences do you see between the ladders in the pictures and real-world ladders?* The ladders in the pictures have only one rail. Although their rungs are parallel, they are not necessarily equally spaced or of the same size.
»   *Based on the examples and nonexamples above, how would you describe a "mathematical" ladder?* A mathematical ladder consists of a line segment (the "rail") with a collection of parallel segments (the "rungs") attached to it at one endpoint along one side of the rail.
»   *What relationships do you see among the angles in the (mathematical) ladders?* The rungs form two sets of congruent angles with rail as shown below.

Next, show students some examples and nonexamples of saws.
Examples of saws:

Nonexamples of saws:

Ask the same types of questions for the saws as you did for the ladders. Elicit the following observations:

>   » The first drawing resembles the blade of a real-world saw, but the others tend to look too "irregular" for this.
>   » All of the drawings consist of line segments joined end to end.
>   » Alternate line segments in the saws are always parallel. In each nonexample, at least one such pair of segments is not parallel.
>   » Neighboring segments of saws always form congruent angles. This is easier to see in the first picture below, but it is true even for saws whose segments are of varying lengths!

Notice that ladders contain one set of parallel segments and two sets of congruent angles, while saws contain two sets of parallel segments and one set of congruent angles.

Explore the relationship between segments and angles in ladders by asking the following questions.

>   » *What happens to the angles in a ladder if you change a line segment so that it is no longer parallel to the others?* The angles associated with that segment are no longer congruent to the others.
>   » *What happens to a rung in a ladder if you change the angle between it and the rail?* It is no longer parallel to the other rungs.

Ask a similar set of questions for saws. Then have students generate four "if/then" sentences that describe their observations.

For "ladder-like" drawings:

1. If rungs are parallel, then the angles between them and the rail are congruent.
2. If angles between the rungs and rail are congruent, then the rungs are parallel.

For "saw-like" drawings:

3. If alternate segments are parallel, then the angles between them are congruent.
4. If angles between segments are congruent, then alternate segments are parallel.

Tell students that, in this exploration, they will spend much of their time looking for or creating ladders and saws within other drawings and using these four observations as a basis for justifying more complex statements. Direct the students' attention to the distinction between statements 1 and 2 (or 3 and 4). Ask them to describe circumstances in which you would use one statement instead of the other. For example, statement 1 is useful if you know that some segments are parallel, and you want to explain why certain angles must be congruent. Statement 2, on the other hand, may be helpful when you know that certain angles are congruent, and you would like to show that one or more pairs of segments is parallel.

STUDENT HANDOUT

Stage 1

1. Draw three triangles—one equilateral, one isosceles, and one scalene. For each triangle, locate and mark the midpoints of all three sides. Connect each pair of midpoints to make three *midsegments* that form an "inner triangle." Make as many observations as you can about the relationships among the sides and angles in your drawings, along with anything else of interest to you. When possible, relate your observations to ladders and saws.

2. Draw a rectangle, parallelogram, trapezoid, kite, and a quadrilateral that has no special name. Make each figure fill most of a page. Mark the midpoint of every side in each figure. Connect the midpoints in order to create a new quadrilateral within each shape. What do all of the "inner quadrilaterals" have in common? Make as many observations as you can, using measurements to justify your statements.

3. Use one or more of your discoveries in Problem #1 to explain why the inner quadrilaterals will always have the common feature(s) that you discovered in Problem #2.

Stage 2

4. Draw and measure the diagonals in each of your quadrilaterals from Problem #1. How do the lengths of the diagonals relate to the perimeters of the inner quadrilaterals? Explain why this happens.

5. Use ladders and saws to answer one or more of the following questions.
   a. Why are opposite angles in a parallelogram congruent?

   b. When you create a copy of a triangle by rotating it around the midpoint of a side, why is the resulting figure a parallelogram? (See Exploration 5, Problem #3.)

   c. Why do triangles have an interior angle sum of 180°?

6. If the lines *l* and *m* are parallel, what is the measure of ∠*CAB* ? What is a general rule for finding the measure of this angle if the given angles have measures of *a*° and *b*? Explain your reasoning for both questions. (Do not assume that the angles in the diagram are drawn accurately.)

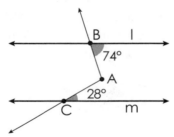

Stage 3

7. Extend Problem #6 by adding another "bend" as shown below. What can you say about the relationship between *m*∠1 and *m*∠2 ? Explain your reasoning.

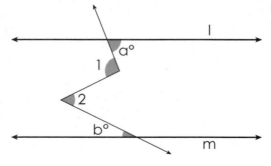

TEACHER'S GUIDE

STAGE 1

Problem #1

1. Draw three triangles—one equilateral, one isosceles, and one scalene. For each triangle, locate and mark the midpoints of all three sides. Connect each pair of midpoints to make three *midsegments* that form an "inner triangle." Make as many observations as you can about the relationships among the sides and angles in your drawings, along with anything else of interest to you. When possible, relate your observations to ladders and saws.

Questions and Conversations for #1

This section contains ideas for conversations, mainly in the form of questions that students may ask or that you may pose to them. Be sure to allow students to do most of the thinking and talking!

» *What types of relationships can you explore among sides?* Consider beginning with side lengths or ways in which sides do or do not intersect.

» *Are there other relationships to look for, aside from those among sides and angles?* You might compare triangles or other figures within the diagrams.

» *Do you see saws or ladders in your drawings? If so, how can you relate them to other observations?* There are many saws and ladders in the diagrams. How can you make them easier to see? What types of measurements are they related to?

Solution for #1

Sample drawings:

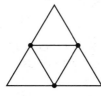 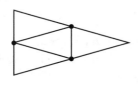 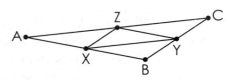

Sample observations for the third drawing. (Analogous observations apply to the first two drawings.)

» The four inner triangles, $\triangle AXZ$, $\triangle ZYC$, $\triangle XBY$, and $\triangle YZX$ appear to be congruent.

> **Teacher's Note.** Remind students that the letters in the triangles' names are ordered so that corresponding vertices are in the same position. For example, in $\triangle AXZ$ and $\triangle ZYC$, vertex A corresponds with vertex Z, X corresponds with Y, and Z corresponds with C.

If the inner triangles are congruent, then you may conclude that:

» $\overline{XY}$ is half the length of $\overline{AC}$. In fact, each *midsegment* is half the length of the side that it does not intersect. This follows from the fact that $\overline{AZ}$ and $\overline{ZC}$ are both congruent to $\overline{XY}$, which in turn is true because the inner triangles are congruent.

» $\overline{XY}$ is parallel to $\overline{AC}$ (written $\overline{XY} \parallel \overline{AC}$). In fact, each midsegment is parallel to the side that it does not intersect. You may verify this using either ladders or saws. The following diagrams show a ladder and a saw in bold.

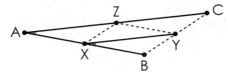 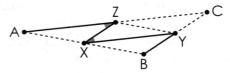

Teacher's Note. Students who are familiar with the concept of *similar* figures may observe that each inner triangle is similar to $\triangle ABC$.

In the ladder on the left, $\overline{XY} \parallel \overline{AC}$ because $\angle CAB \cong \angle YXB$. The saw on the right shows that $\overline{XY} \parallel \overline{AC}$ because $\angle AZX \cong \angle ZXY$. In both cases, the given angles are congruent because they are corresponding angles in congruent triangles.

Problem #2

2. Draw a rectangle, parallelogram, trapezoid, kite, and a quadrilateral that has no special name. Make each figure fill most of a page. Mark the midpoint of every side in each figure. Connect the midpoints in order to create a new quadrilateral within each shape. What do all of the "inner quadrilaterals" have in common? Make as many observations as you can, using measurements to justify your statements.

Questions and Conversations for #2

» *Why are you encouraged to make your drawings large enough to fill most of a page?* You can make better visual observations and more accurate measurements when you work with large drawings.

» *How do your measurements relate to the categories to which the shapes belong?* Use your knowledge of the properties of different types of quadrilaterals.

» *Do any of your inner quadrilaterals belong to more than one category?* Yes. Keep this is mind as you search for something common to all of them.

Solution for #2

Sample drawings.

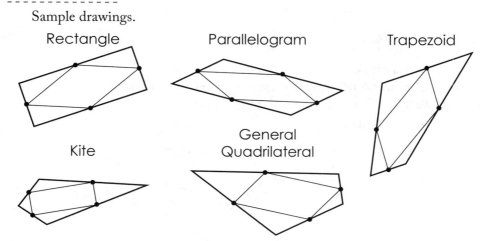

The key characteristic shared by all of the inner quadrilaterals is that they are parallelograms. Students may verify this by measuring to determine that opposite sides and/or opposite angles of these figures are congruent. (This does not *prove* that the inner quadrilaterals are parallelograms, because our definition of a parallelogram is stated in terms of parallel sides, not side lengths or angles. Instead, students are making use of familiar *properties* of parallelograms.)

> **Teacher's Note.** Students may notice that some of the inner quadrilaterals are rectangles or rhombuses. If they are not accustomed to thinking of these figures as special types of parallelograms, they may not notice that all of the inner quadrilaterals are parallelograms. This is an excellent opportunity to talk about geometric figures belonging to multiple categories.

Problem #3

3. Use one or more of your discoveries in Problem #1 to explain why the inner quadrilaterals will always have the common feature(s) that you discovered in Problem #2.

Questions and Conversations for #3

» *How can you apply what you learned from Problem #1 to this situation?* It may help to add *auxiliary* features to your diagram (i.e., to insert additional lines or segments in order to view the drawing in a new and helpful way).

» *How is this question different from Problem #2, in which you were asked to justify your conclusion that the inner quadrilaterals are parallelograms?* In Problem #2, you based your conclusion on properties of parallelograms observed in sample measurements—not on the definition of a parallelogram. To explain why the inner quadrilateral must always be a parallelo-gram, you need a reason that is based directly on this definition.

Solution for #3

Draw a diagonal of one of the original quadrilaterals. This forms two triangles whose midsegments are opposite sides of the inner quadrilateral. For example, see the dotted segment $\overline{AB}$ below. $\overline{MN}$ is a midsegment of $\triangle ABC$, and $\overline{XY}$ is a midsegment of $\triangle ABD$.

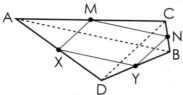

According to the results of Problem #1, $\overline{AB} \parallel \overline{MN}$ and $\overline{AB} \parallel \overline{XY}$. Because $\overline{MN}$ and $\overline{XY}$ are parallel to the same segment, they are parallel to each other. You can use the same type of reasoning with the diagonal $\overline{CD}$ to show that $\overline{MX} \parallel \overline{NY}$. Because both pairs of opposite sides of quadrilateral $MNYX$ are parallel, it is a parallelogram.

STAGE 2

Problem #4

4. Draw and measure the diagonals in each of your quadrilaterals from Problem #1. How do the lengths of the diagonals relate to the perimeters of the inner quadrilaterals? Explain why this happens.

Questions and Conversations for #4

» *What can you try if you are having trouble predicting a relationship?* Organize the data for the diagonals and perimeters. Look for patterns.
» *Might it help to look at a smaller portion of the data for each quadrilateral?* Yes. Look at one diagonal (and one pair of opposite sides).

Solution for #4

The sum of the lengths of the diagonals of a quadrilateral is always equal to the perimeter of its inner parallelogram. For example, in the diagram for Problem #3, $\overline{MN}$ and $\overline{XY}$ are each half the length of $\overline{AB}$. Therefore, the sum of their lengths is equal to the length of $\overline{AB}$:

$$AB = MN + XY$$

A similar comment applies to the diagonal $\overline{CD}$:

$$CD = MX + NY$$

The desired conclusion follows from combining these measurements:

$$AB + CD = MN + XY + MX + NY$$

The left side of the equation represents the sum of the lengths of the diagonals, and the right side is the perimeter of the inner parallelogram.

Problem #5

5. Use ladders and saws to answer one or more of the following questions.
   a. Why are opposite angles in a parallelogram congruent?
   b. When you create a copy of a triangle by rotating it around the midpoint of a side, why is the resulting figure a parallelogram? (See Exploration 5, Problem #3.)
   c. Why do triangles have an interior angle sum of 180°?

Questions and Conversations for #5

» *Part a: What happens if you extend the sides of a parallelogram?* It enables you to see the parallelogram as part of a larger pattern, and it creates a set of angles that may be helpful to work with.

» *Part a: Do you see any saws or ladders in the picture with the extended sides?* There are plenty of them. Experiment to find those that are helpful.

» *Part b: How might you use saws or ladders differently in this question than you did in part a?* In part *a*, you knew some parallel sides, and you were looking for congruent angles. In this question, the situation is reversed.

» *Part c: How can you deal with the fact that a triangle has no saws or ladders?* Draw one or more auxiliary lines in order to create saws or ladders.

» *Part c: Without measuring, how might you draw three angles whose sum is obviously 180°?* Place them next to each other so that they form a straight angle.

Solution for #5

Sample solution for a. Because the figure is a parallelogram, its opposite sides are parallel. Therefore, you may use ladders, saws, or both to find congruent angles. The following example shows one way to explain why $\angle 1 \cong \angle 3$. (Notice how it helps to extend the sides of the parallelogram.)

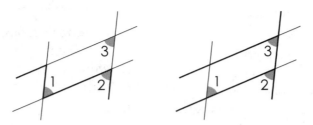

The saw in the left drawing shows that $\angle 1 \cong \angle 2$. In the drawing on the right, the ladder shows that $\angle 2 \cong \angle 3$. Because $\angle 1$ and $\angle 3$ are both congruent to $\angle 2$, they are congruent to each other.

Sample solution for b. In this case, you know that certain angles are congruent, because the triangles to which they belong are congruent. You may use saws to find parallel segments. (Some students may develop other lines of reasoning using ladders.)

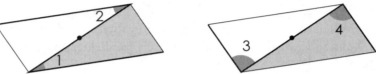

In these drawings, the shaded triangle is rotated about the marked midpoint to produce a congruent (unshaded) triangle above it. The marked angles within each drawing are congruent because they are corresponding angles in the congruent triangles. Because $\angle 1 \cong \angle 2$ in the first drawing, the saw shows that the long sides of the quadrilateral are parallel. The saw in the second drawing makes use of $\angle 3$ and $\angle 4$ to show that the left and right sides of the quadrilateral are parallel. Because both pairs of opposite sides are parallel, the quadrilateral is a parallelogram.

Sample solution for c. Because a triangle does not contain ladders or saws, you need to draw one or more *auxiliary* lines (see Questions and Conversations for #3) in order to create a picture for which you can use this type of reasoning. For example, you may draw a line parallel to a base through the opposite vertex.

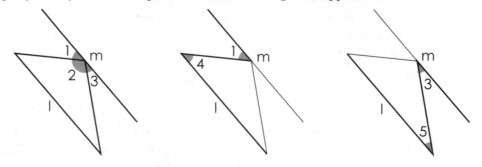

The first drawing shows an auxiliary line m that is parallel to the side labeled l. This drawing shows that:

$$m\angle 1 + m\angle 2 + m\angle 3 = 180°$$

Only $\angle 2$ in this equation is an interior angle of the triangle. However, you may use saws to connect $\angle 1$ and $\angle 3$ to angles that are within the triangle. In the middle drawing, the saw shows that $\angle 1 \cong \angle 4$. The saw in the picture on the right shows that $\angle 3 \cong \angle 5$. If you replace the angles in the equation with the ones to which they are congruent, you obtain:

$$m\angle 4 + m\angle 2 + m\angle 5 = 180°$$

The left side of the equation is the sum of the interior angles of the triangle!

Problem #6

6. If the lines l and m are parallel, what is the measure of $\angle CAB$? What is a general rule for finding the measure of this angle if the given angles have measures of $a°$ and $b°$? Explain your reasoning for both questions. (Do not assume that the angles in the diagram are drawn accurately.)

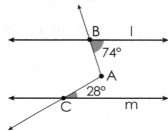

Questions and Conversations for #6

» *Are you allowed to draw your own accurate picture and make measurements?* Yes, especially if you are unsure how else to begin. Consider making multiple pictures with different angle measurements. Then look for patterns.

» *Does the diagram contain any ladders or saws?* Not quite. Try to create some.

» *Do you see any vertical or supplementary angles?* The diagram contains both. You may create more by drawing auxiliary lines.

» *Is it okay to make use of interior angle sums for polygons in the reasoning process?* Yes.

» *Once you think you've found a pattern or rule to predict $m\angle CAB$, how can you be certain that it will always work?* Try to understand what causes the pattern. Consider using variables. Or imagine making changes to the drawing and observing their effects.

Solution for #6

The measure of $\angle CAB$ is $102°$. Some students may discover this by making an accurate drawing and then measuring the angle. Because $102°$ is the sum of $74°$ and $28°$, they may conjecture that $m\angle CAB$ is always the sum of the measures of the marked angles. Other students may begin by using reasoning strategies such as those below.

Sample strategy 1. Although this strategy involves a relatively long chain of reasoning, students often discover it. Extend one of the rays in the opposite direction to form a triangle. For instance, in the drawing below, $\overline{AP}$ creates $\triangle BAP$.

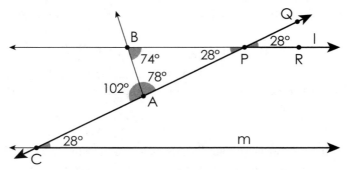

From here, the reasoning process may follow many routes. For example:
» The ladder in bold shows that $m\angle QPR = 28°$.
» $m\angle APB = 28°$, because $\angle QPR$ and $\angle APB$ are vertical angles.
» $m\angle BAP = (180 - (74 + 28))° = 78°$, because the sum of the interior angles in $\triangle BAP$ is 180°.
» $m\angle CAB = (180 - 78)° = 102°$ because $\angle CAB$ and $\angle BAP$ are supplementary angles.

The first two steps could be reduced to one step by using a saw instead of a ladder. (Can you see how?)

To see that $m\angle CAB$ is always the sum of the given angles, you may replace the measured values in the diagram with variables, and use the same reasoning process.

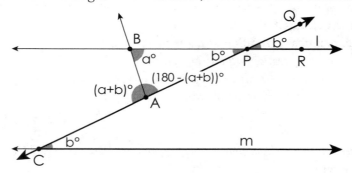

» The ladder in bold shows that $m\angle QPR = b°$.
» $m\angle APB = b°$, because $\angle QPR$ and $\angle APB$ are vertical angles.
» $m\angle BAP = (180 - (a + b))°$, because the sum of interior angles in $\triangle BAP$ is 180°.

» $m\angle CAB = 180° - \left(180 - (a + b)\right)° = (a + b)°$, because $\angle CAB$ and $\angle BAP$ are supplementary angles.

Students should not be deterred by the apparent complexity of the expression $180° - \left(180 - (a + b)\right)°$ in the final step, even if they do not know algebraic processes. Instead, they may ask a question such as, "What must you add to $180 - (a + b)$ in order to obtain 180?"

Sample strategy 2. Draw an auxiliary line through A parallel to lines l and m.

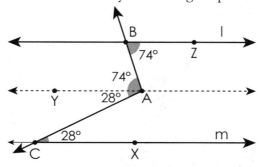

The auxiliary line creates saws from X to C to A to Y and from Y to A to B to Z. The saws show that $m\angle CAY = 28°$ and $m\angle BAY = 74°$ respectively. Therefore, $m\angle CAB = m\angle CAY + m\angle BAY = 28° + 74° = 102°$.

This strategy makes it intuitively clear that $m\angle CAB$ will always be equal to the sum of the given angles. Many students who use this strategy describe their thinking without referring to auxiliary lines. Instead, they imagine sliding the lines l and m together toward the point A, which helps them "see" $\angle CAB$ as being comprised of the given angles.

Other strategies. You may build other strategies based on auxiliary lines like those below. Encourage students to explore these situations, even if they did not think of them!

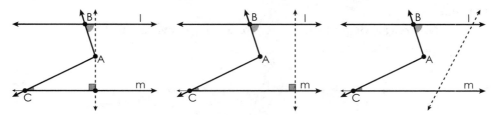

STAGE 3

Problem #7

7. Extend Problem #6 by adding another "bend" as shown below. What can you say about the relationship between $m\angle 1$ and $m\angle 2$? Explain your reasoning.

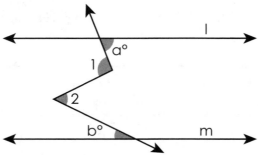

Questions and Conversations for #7

»   *How can you get started if you don't know where to draw the auxiliary line(s)?* As before, you could begin by making accurate drawings using different values of *a* and *b,* measuring other angles, organizing your data, and looking for patterns. Or you could experiment with different auxiliary lines to create ladders and saws until you find something helpful.

Solution for #7

The difference between the measures of $\angle 1$ and $\angle 2$ is constant. Specifically, $m\angle 1 - m\angle 2$ is always equal to $(a - b)°$. Some students may discover this by drawing diagrams using various values for *a* and *b,* making measurements, organizing the data, and looking for patterns. Others may begin by using algebra to explore the general situation.

Sample strategy. Begin by drawing two auxiliary lines parallel to lines *l* and *m* through the vertices at the "bends."

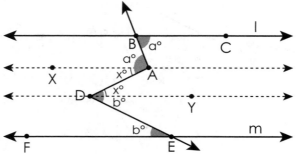

The saws from *C* to *B* to *A* to *X* and from *F* to *E* to *D* to *Y* show that $m\angle BAX = a°$ and $m\angle EDY = b°$ respectively. Although the value of *x* in the diagram may vary, $m\angle YDA$ is always equal to $m\angle DAX$. (See the saw from X to A to D to Y.)

Comparing the diagram in the original problem to the picture above, $m\angle 1 = (a+x)^\circ$ and $m\angle 2 = (b+x)^\circ$. Therefore:

$$m\angle 1 - m\angle 2 = (a+x)^\circ - (b+x)^\circ$$
$$= (a-b)^\circ$$

> **Teacher's Note.** If students do not understand how to obtain the expression $a-b$ from $(a+x)-(b+x)$, ask them to think about how the value of a subtraction calculation is affected when you add the same number to both of the numbers that you are subtracting.

Some students may develop a strategy based on the drawing below. For example, they may begin by using a saw to show that the two angles marked z° are congruent.

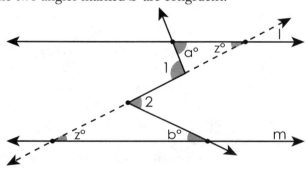

Students may wonder what happens if you slide the points B and E into different positions on their lines or if you move points A and D so that D lies above A in the diagram as illustrated below. Encourage them to explore these situations! Do the conclusions remain the same? Do the thinking strategies change?

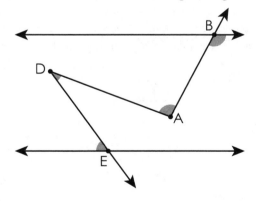

131

WRAP UP

Share Strategies

Give students an opportunity to compare and contrast their thinking strategies, identifying advantages and disadvantages of each.

Summarize

Answer any remaining questions that students have. Summarize and extend key ideas. Some suggestions are shown below.

» Auxiliary lines may be useful when you do not see helpful features such as ladders, saws, or polygons in the original drawing.

» *Saws* and *ladders* are not standard mathematical terminology. Students who would like to learn the traditional vocabulary may be interested in reading about parallel lines and *transversals*.

» Ask students if they used any assumptions other than what they know of saws and ladders to answer Problem #1. Many of them may have assumed that the four small triangles are congruent. If so, ask them how they know this. They typically realize that they based their conclusion on the general appearance of the triangles and that they would have to make additional assumptions (beyond saws and ladders) to justify it in any other way. This may lead to a "chicken or egg" discussion about where the reasoning process begins! At this point, you may tell students that mathematicians do try to build all of their logical arguments on a foundation of the fewest possible "obvious" facts. These ideas are very important in mathematics and students will explore them further in more advanced courses.

» Statements (such as those numbered 1 and 2 [or 3 and 4] in the introduction) whose *if* and *then* components are interchanged are called *converses*. Ask students if they can generate examples of true statements whose converses are false. (There are countless examples. For example, the statement "If a figure is a quadrilateral, then it is a polygon" is true. However, the converse "If a figure is a polygon, then it is a quadrilateral" is false.) Understanding converses is an important part of learning to reason mathematically.

Further Exploration

Ask students to think of new questions to ask or ways to extend this exploration. Here are some possibilities:

» Is the inner quadrilateral of a rectangle always a rhombus? Is the inner quadrilateral of a kite always a rectangle? Explain.

» In Problem #2, make a conjecture about the relationship between the area of each quadrilateral and its inner quadrilateral. Explain why this happens.

» If you have not done so already, extend the results of Problems #2 through #4 by exploring concave kites.

» Extend Problem #5 by using ladders and saws to justify other properties of triangles or quadrilaterals.

» Extend Problem #7 by continuing to add more "bends" or changing the diagram in other ways.

» Explore Problem #5c further by doing some research on the *Parallel Postulate* and *Non-Euclidean Geometry*.

Exploration 8

Designing Nets

INTRODUCTION

Materials

- » Compass, protractor, ruler, and calculator
- » Poster board, scissors, and tape
- » Sand or popcorn to test the volume of a cone (optional)

Prior Knowledge

- » Know and understand a formula for the volume of a rectangular prism.
- » Know and be able to apply the Pythagorean Theorem.
- » Find areas of triangles, quadrilaterals, and circles.
- » Vocabulary: *prism, pyramid, cone, base, height, face, congruent, dimensions, apex* (of a pyramid or cone), *volume, surface area*

> **Teacher's Note.** This exploration works best if students do not know formulas for volumes or surface areas of pyramids or cones. I generally use it over an extended period of time, interspersing the problems with some routine practice using the formulas (once students have discovered them).

Learning Goals

- » Understand why the formula $V = B \cdot h$ makes sense for the volume of a prism.
- » Practice spatial visualization skills.
- » Use *nets* to develop and justify formulas for volumes and lateral surface areas of right pyramids and cones.
- » Investigate relationships between volume and surface area.
- » Attend to precision of numbers when doing calculations and reporting results.
- » Communicate complex mathematical ideas clearly.
- » Persist in solving challenging problems.

Launching the Exploration

Motivation and purpose. To students: A *net* is a two-dimensional figure that may be folded to form a three-dimensional figure. Nets can help you develop spatial visualization skills. In this activity, you will also use them to create and apply your own methods for calculating volumes and surface areas of pyramids and cones.

Understanding the problem. Skim the entire exploration with students. In Stage 1, they investigate connections between volumes and surface areas of rectangular prisms in order to prepare for similar work with pyramids and cones in Problems #4 and #6. In the remaining problems (#5 and #7), students apply their new knowledge to solve problems and to create and justify formulas.

Use the following three demonstrations with students to investigate the effects of moving one side or vertex of a figure while leaving its base fixed.

1. What do you notice about the areas of these triangles? Why does this happen? What can you say about the segments that intersect the triangles? What about the perimeters of the triangles?

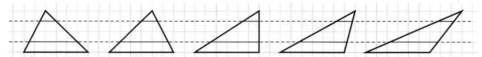

As you drag the top vertex to the right (but not up or down), the triangles maintain the same area, because their bases and heights remain the same. When you draw a line parallel to a base through all of the triangles, it intersects the triangles in segments of the same length. As the vertex moves, imagine a collection of congruent segments at each height moving along with it. This may help you visualize why the area does not change. The perimeters, however, do change because two of the sides become longer or shorter depending on the position of the vertex.

2. Take a deck of cards and "tilt" it so that it approximates the form of an *oblique* rectangular prism—a prism for which the *lateral* faces (the "nonbase" faces) are not all perpendicular to the base. What can you say about the volumes of the prisms? Why does this happen? What can you say about the *cross sections* (the surfaces formed by slicing the prisms) at each height? Does the surface area remain the same when you tilt the prism?

Both prisms have the same volume, because the area of the cross section at every height (the area of a card) is equal. The heights of the prisms are also equal, as are the areas of the bases. (Note that the height is still measured perpendicular to the base.) However, the total area of the lateral faces of the prism on the right is greater because some of the edges of the tilted faces are "stretched."

3. Take an empty cereal box and break open and cut off the flaps at the ends. Tilt it so that it forms an oblique rectangular prism. What happens to the volume of the prism? How is this situation different than the deck of cards?

    End view of the cereal box:

    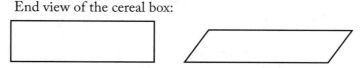

    The volume changes in this case because, although both bases (the top and bottom faces) have the same area, the heights of the prisms are now different. You can see this clearly by continuing to tilt the box until it is completely flat. Some students may observe that the surface area of the box remains constant. (No cardboard is created or destroyed in the process of tilting the box!)

Based on the three examples above, lead a discussion in which you guide students to the following conclusions:
   » The volume of a prism depends upon the area of its base and its height.
   » If a prism is "tilted" to form an oblique prism with the same base area and height, the volume remains unchanged. However, the lateral surface area may change.

Students should find it reasonable that similar conclusions apply to cylinders, pyramids, and cones. Encourage them to visualize and discuss these situations.

STUDENT HANDOUT

Stage 1

1. A square prism has a volume of 192 cm³. A second square prism has a total surface area of 224 cm² and a lateral surface area of 192 cm². Tell which statement is true and explain your reasoning.
   » The two prisms must be congruent.
   » The two prisms might be congruent.
   » The two prisms cannot be congruent.

2. Create a new prism from the first prism in Problem #1 by dividing the sides of its bases into three congruent segments and cutting a triangle from each vertex as shown below. Is it possible to calculate the volume of the resulting octagonal prism? If it is possible, do so, and explain your thinking. If not, explain why.

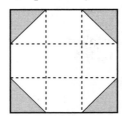

3. Change the second prism from Problem #1 into an octagonal prism in the same way as in Problem #2. Is it possible to calculate the surface area of the octagonal prism? If it is possible, do so, and show your thinking. If not, explain why.

4. Use the square on the final page of this handout [p. 140] to create a net for a square pyramid whose apex is directly above a vertex of the base. Cut it out, fold it, and tape it to form the pyramid. Make just enough identical copies of this pyramid so that you are able to put them together to form a rectangular prism. Make the necessary measurements and find the surface area (cm²) and volume (cm³) of the pyramid. Explain your thinking.

5. At the time of its construction, the Great Pyramid of Giza had a volume of about 9.16×10^7 ft³ and a height of about 481 ft. What was its approximate lateral surface area? Use diagrams, calculations, and words to explain your thinking.

Stage 3

6. Choose a radius and a volume for a cone that you would like to build. Create a net for the cone and construct it. Explain how to find its lateral surface area.

7. Predict and test a formula for the lateral surface area of a cone. If your formula does not work, continue predicting and testing until you find one that does. Prove that your formula applies to all cones.

NET FOR PROBLEM #4

TEACHER'S GUIDE

STAGE 1

Problem #1

1. A square prism has a volume of 192 cm³. A second square prism has a total surface area of 224 cm² and a lateral surface area of 192 cm². Tell which statement is true and explain your reasoning.
   - » The two prisms must be congruent.
   - » The two prisms might be congruent.
   - » The two prisms cannot be congruent.

Questions and Conversations for #1

This section contains ideas for conversations, mainly in the form of questions that students may ask or that you may pose to them. Be sure to allow students to do most of the thinking and talking!

- » *What does* lateral *surface area mean?* A lateral surface (or face) is a surface that is not a base.
- » *How do you get started?* There are many approaches. If you find it difficult to think of both prisms at once, choose one of them to begin with.
- » *Are there choices for the values of the dimensions of the prisms?* It may depend upon the prism. If you do have options, choose numbers for the measurements, and try to calculate the effects.

Solution for #1

The two prisms might be congruent.

Sample strategy. The total surface area of the square bases of the second prism is $224 - 192 = 32$ cm². Therefore, each of its bases has an area of $\frac{32}{2} = 16$ cm². The length of each side of a base is thus $\sqrt{16} = 4$ cm. Because there are four congruent lateral faces, and the lateral surface area of this prism is 192 cm², the area of each lateral face is $\frac{192}{4} = 48$ cm². Because the length of one side of each lateral face equals the height of the prism, and the other is a side of the base (4 cm), the height of the prism must be $\frac{48}{4} = 12$ cm. The volume of the prism is therefore $V = l \cdot w \cdot h = 4 \cdot 4 \cdot 12 = 192$ cm³.

Because the second prism must have a volume of 192 cm³, some students may conclude that the two prisms must be congruent. However, there are many other square prisms with volumes of 192 cm³ that do not fit the conditions for the second prism. For example, a prism with dimensions $l = 3$ cm, $w = 3$ cm, and $h = 21\frac{1}{3}$ cm has the desired volume, but its total surface area (274 cm²) and lateral surface area (256 cm²) are different than those of the second prism.

Problem #2

2. Create a new prism from the first prism in Problem #1 by dividing the sides of its bases into three congruent segments and cutting a triangle from each vertex as shown below. Is it possible to calculate the volume of the resulting octagonal prism? If it is possible, do so, and explain your thinking. If not, explain why.

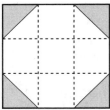

Questions and Conversations for #2

» *How can you find the volume when you don't know the dimensions of the original prism?* If you are not sure, experiment with a variety of choices for l, w, and h. Just be sure that they result in a volume of 192 cm³. Think about how to find the volumes of the triangular prisms that you remove from the original prism.

» *What happens to the area of a base when you cut off the corners?* Although you cannot predict the *amount* by which it decreases (because that depends on the original size of the square), you can determine the *factor* by which it changes. In other words, you can determine what fraction it is of the original area.

» *How is the volume affected when you change the bases?* This is a key question. Investigate by visualizing simpler changes to the bases. Try different values for the measurements.

Solution for #2

Some students may believe that it is impossible to determine the volume of the octagonal prism, because there are many possible values for the length, width, and height of the square prism. However, if they experiment with a variety of measurements for these dimensions, they will find that the volume always comes out to $149\frac{1}{3}$ cm³!

The diagram for this problem shows that, regardless of the side lengths, the area of the octagonal base is always $\frac{7}{9}$ of the area of the square. Therefore, the volume of the prism will also be $\frac{7}{9}$ of its original volume:

$$192 \cdot \frac{7}{9} = 149\frac{1}{3} \text{ cm}^3, \text{ or } 192 \div 9 \cdot 7 = 149.\overline{3} \text{ cm}^3$$

If students feel uncomfortable with this idea, ask them to think about what happens to the volume of the prism when they make the area of a base one half as large.

The ideas in this problem suggest that the area of the base is an important consideration in determining the volume of the prism. Suppose that B represents the area of a base. For rectangular prisms, you may rewrite the formula $V = l \cdot w \cdot h$ as $V = B \cdot h$. What happens if you apply the formula $V = B \cdot h$ to prisms with other bases—for example, to this problem in which the base becomes $\frac{7}{9}$ as large? If V, B, and h represent values for the square prism, then the volume of the octagonal prism is:

$$\left(\frac{7}{9} \cdot B\right) \cdot h = \frac{7}{9} \cdot (B \cdot h) = \frac{7}{9} \cdot V$$

Using the associative property of multiplication to regroup the numbers and variables shows that as the area of the base becomes $\frac{7}{9}$ as large, the volume of the octagonal prism does the same. The formula $V = B \cdot h$ produces the predicted effect on the volume!

This formula applies to cylinders as well. Although prisms are usually defined to have polygons for bases, cylinders possess most of the same key features as prisms (including two congruent, parallel bases). In many ways, it makes sense to think of a cylinder as a "circular prism!"

Problem #3

3. Change the second prism from Problem #1 into an octagonal prism in the same way as in Problem #2. Is it possible to calculate the surface area of the octagonal prism? If it is possible, do so, and show your thinking. If not, explain why.

Questions and Conversations for #3

» *How many different square prisms have the surface area measurements given in Problem #1?* There is only one square prism that has these measurements.

» *Do you expect the surface area to be greater or less than that of the square prism? Why?* It will be less for two reasons. First, the area of each base decreases.

Second, the area of each grey lateral face (see the picture in the solution) is less than the area of the two parts of the faces that it replaces.

» *Are all sides of the octagonal bases the same length?* No. Which ones are longer? Why? How can you find their lengths?

» *After students have found the surface area: Could you have found the surface area of the octagonal prism if you had been given only the total surface area of the square prism?* No. Try it with different prisms having the given surface area. The areas of the bases and the areas of the lateral faces change by different amounts.

Solution for #3

There is only one square prism having the given measurements. Consequently, it is possible to calculate the surface area of the octagonal prism directly. The result is:

$$88\frac{8}{9} + 64 \cdot \sqrt{2} \approx 179.4 \ \text{cm}^2$$

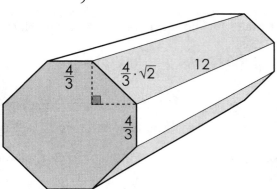

Sample strategy: The total area of the square bases was 32 cm². The area of the bases of the octagonal prism is $\frac{7}{9}$ of this:

$$32 \cdot \frac{7}{9} = 24\frac{8}{9} \ \text{cm}^2$$

There are four lateral faces like the two shown in white, each of which has an area of $12 \cdot \frac{4}{3} = 16 \ \text{cm}^2$. Their total area is $16 \cdot 4 = 64 \ \text{cm}^2$.

To find the areas of the light grey lateral surfaces, begin by using the Pythagorean Theorem with the triangle shown below (and in the picture of the prism above).

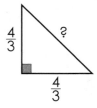

The length of its hypotenuse forms a side of these lateral faces. Depending on how students perform the calculations, they may express their answers in a number of equivalent ways, including $\sqrt{\frac{32}{9}}$, $\frac{\sqrt{32}}{3}$, $\frac{4 \cdot \sqrt{2}}{3}$, and $\frac{4}{3} \cdot \sqrt{2}$. Because each of these faces has a 12 cm side, you may express its area as

$$\left(\frac{4}{3} \cdot \sqrt{2}\right) \cdot 12 = \left(\frac{4}{3} \cdot 12\right) \cdot \sqrt{2} = 16 \cdot \sqrt{2}.$$

The total area of all four faces is

$$4 \cdot \left(16 \cdot \sqrt{2}\right) = (4 \cdot 16) \cdot \sqrt{2} = 64 \cdot \sqrt{2}.$$

The surface area of the octagonal prism is the sum of the areas of the bases and both types of lateral faces:

$$24\frac{8}{9} + 64 + 64 \cdot \sqrt{2} = 88\frac{8}{9} + 64 \cdot \sqrt{2} \approx 179.4 \text{ cm}^2$$

> **Teacher's Note.** If students have done many of their calculations by writing numbers in decimal form, encourage them to solve the problem again without using a calculator until the end. Challenge them to keep their results in exact form (in terms of fractions and square roots) for as long as possible. Ask them to think about why this may improve the accuracy of their results. Incidentally, it also provides excellent practice with manipulating expressions involving fractions and square roots!

STAGE 2

Problem #4

4. Use the square on the final page of this handout [p. 140] to create a net for a square pyramid whose apex is directly above a vertex of the base. Cut it out, fold it, and tape it to form the pyramid. Make just enough identical copies of this pyramid so that you are able to put them together to form a rectangular prism. Make the necessary measurements and find the surface area (cm²) and volume (cm³) of the pyramid. Explain your thinking.

Questions and Conversations for #4

> » *What can you do if you have trouble visualizing how the net will fold together?* Make some "practice" nets. Don't worry about the exact measurements yet, but pay close attention to which parts of the net must "match" when you fold them. If the net doesn't fold properly, look back to see if you need to change the orientation of the triangles or the lengths of their sides.

> **Teacher's Note.** Make a completed copy of a pyramid to show to students who are having trouble visualizing what it will look like. Ask them to imagine unfolding it.

Solution for #4

Students often choose to make the height of the pyramid equal to the side lengths of the square. The following picture shows the net for this situation with the measurements rounded to the nearest millimeter.

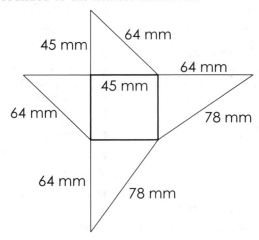

After cutting, folding, and taping, a completed pyramid will look like this:

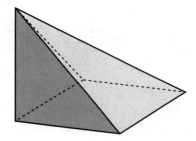

Students should find that three of these pyramids are needed to create a cube. (If they choose a different height, the rectangular prism will not be a cube, but they will still need three pyramids.) Consequently, the volume of the pyramid is one-third of the volume of the cube.

$$V = \frac{1}{3} \cdot 45 \cdot 45 \cdot 45 = \frac{1}{3} \cdot 45^3 = 30{,}375 \text{ mm}^3$$

If students believe that their measurements could be off by as little as 1 mm, it might make sense to round this answer to about 30,000 mm³, because a change of 1 mm in the length of each side makes quite a difference in the result: $\frac{1}{3} \cdot 44 \cdot 44 \cdot 44 \approx 28{,}395$ and $\frac{1}{3} \cdot 46 \cdot 46 \cdot 46 \approx 32{,}445$! Keep in mind how small a cubic millimeter is!

The surface of the pyramid consists of the base, two triangles that are half the area of the base, and two larger triangles. To find the surface area, begin by noticing that the hypotenuse of the middle triangle is the same length as the longer leg of the triangle on the right. (If necessary, look back to see how the net fits together.) Use the Pythagorean Theorem to find this length to be $45 \cdot \sqrt{2} \approx 64$ mm.

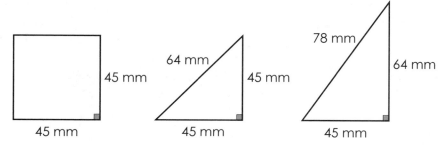

The surface area is:

(area of the square) + 2 • (area of the first triangle) + 2 • (area of the second triangle) =

$$45^2 + 2 \cdot \left(\frac{1}{2} \cdot 45^2 \right) + 2 \cdot \left(\frac{1}{2} \cdot 45 \cdot 45\sqrt{2} \right) \approx 6914 \text{ mm}^2$$

Notice that the result may be more accurate if you use $45 \cdot \sqrt{2}$ as the height of the triangle on the right. (If you round this to 64 in your calculations, you get a value of 6930 mm².) However, given the precision of the measurements, it probably makes sense to round the answer to about 6900 mm² anyway.

Problem #5

5. At the time of its construction, the Great Pyramid of Giza had a volume of about 9.16×10^7 ft³ and a height of about 481 ft. What was its approximate lateral surface area? Use diagrams, calculations, and words to explain your thinking.

> **Teacher's Notes.** (1) The Great Pyramid is a *right* pyramid. That is, its apex is directly above the center of its base. This makes a difference for the surface area! (2) If students are unfamiliar with scientific notation, you may tell them that 9.16×10^7 is equal to 91,600,000.

Questions and Conversations for #5

» *How is the height of the pyramid measured?* It must be measured perpendicular to its base.

» *Based on your experience in Problem #4, how would you find the volume of a pyramid like this?* You find the volume by multiplying the area of the base by the height of the pyramid and taking one third of the result. Does it matter that the vertex is above the center of the base instead of one of its vertices? (See "Understanding the Problem" in the Introduction to the exploration.)

» *How can you use a formula to describe this process?* $V = \frac{1}{3} \cdot B \cdot h$

» *How does the height of a triangle on a lateral face compare to the height of the pyramid?* The height of the triangles is greater than the height of the pyramid. Draw a picture to show why!

» *How can you use the height of the pyramid to find its slant height?* Visualize right triangles within the pyramid!

Solution for #5

The lateral surface area was approximately 9.25×10^5 or 925,000 ft².

Sample strategy. Some students may reason verbally. Because the volume of the pyramid is $\frac{1}{3}$ the product of the area of the base and the height, 3 times the volume is equal to the area of the base times the height. Therefore, the area of the base is 3 times the volume divided by the height. Other students may represent their thinking algebraically:

$$V = \frac{1}{3} \cdot B \cdot h$$

$$3 \cdot V = B \cdot h$$

$$B = \frac{3 \cdot v}{h}$$

The length of a side of the base, a, is the square root of this quantity.

$$a = \sqrt{\frac{3 \cdot V}{h}} = \sqrt{\frac{3 \cdot \left(9.16 \times 10^7\right)}{481}} \approx 755.8504 \text{ ft}$$

This diagram shows that the height (l) of a triangle on the pyramid's surface is greater than the height (h) of the pyramid itself.

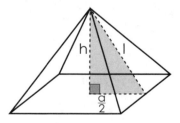

Use the Pythagorean Theorem with $\frac{a}{2} \approx \frac{755.8504}{2} \approx 377.9252$ and $h = 481$ to find the value of l:

$$377.9252^2 + 481^2 \approx l^2$$

This results in a value of about 611.7095 ft for l. To find the lateral surface area, calculate the area of one of the triangles and multiply it by 4:

$$\left(\frac{1}{2} \cdot 755.8504 \cdot 611.7095\right) \cdot 4 \approx 925,000 \text{ ft}^2$$

Some students will not take the most direct route to find the slant height. For example, they may first find the length of an edge (labeled x in the picture) by using the right triangle shown below.

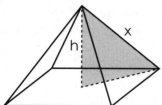

Can you see how they might carry out the entire process with multiple applications of the Pythagorean Theorem?

STAGE 3

Problem #6

6. Choose a radius and a volume for a cone that you would like to build. Create a net for the cone and construct it. Explain how to find its lateral surface area.

Questions and Conversations for #6

» *Where do you see cones in the world?* They appear in many places: party hats, ice cream cones, drinking cups, traffic cones, funnels (with the tip cut off!), crayon tips, megaphones, audio speakers, etc.

» *How do you calculate the volume of a cone? Why?* Because a cone looks like a pyramid with a circular base, students may conjecture (correctly) that you find its volume using the same process as for a pyramid—by taking one third the product of the area of the base and the height. (After all, a regular polygon with many sides looks a lot like a circle!) In symbols: $V = \frac{1}{3} \cdot B \cdot h$.

However, the area of the base is now given by the expression $\pi \cdot r^2$.

» *What should be the radius of the base?* Students often settle on a value between 3 and 6 cm.

Teacher's Note. This is a good opportunity to discuss metric units for volume. Tell students that 1 *liter* (L) is equal to 1000 cm³, and show them what this looks like. Ask them how to name 1 cm³ in terms of liters. (Answer: 0.001 L or 1 *milliliter* (mL)).

» *What should be the volume of the cone?* Students tend to guess small values resulting in cones that look very "flat." As they make suggestions, ask them to calculate the resulting height. If they do not like the result, encourage them to think about what change in the volume would be needed to make the height come out to a value that they like better. (If you would like students' choices to match those in the Solution for #6, steer them toward the values $r = 5$ cm and $V = 200$ cm³.)

» *What does a net for the lateral surface look like?* Allow students to suggest possibilities. Someone usually suggests a "pie shape" (a *sector* of a circle). Students may verify that this works by cutting one out and folding it.

» *Which will be larger—the base of the cone or the circle used for the lateral surface's net? Why?* Students generally predict correctly that the circle used for the lateral surface must be larger so that when you take a sector of it and fold it, it will still be large enough to cover the base.

» *How can you determine the radius of the larger circle?* If students are not sure, have them make an example to cut and fold. Where does the radius of the larger circle appear on the lateral surface?

» *How can you determine the size of the sector needed to make the base and the lateral surface fit together properly?* By visualizing or experimenting, students discover that the length of the arc on the sector must equal the circumference of the base.

» *What is the central angle needed to produce the correct arc length?* If students struggle with this question (as they often do), ask them to suppose that the radius of the circle for the lateral surface has twice the circumference of the base. What is the central angle in this case? (The arc length of the sector must be half the circumference of the larger circle. Therefore, the central angle must be 180° [half of 360°]). How can you generalize this idea for other ratios between the circumferences?

Solution for #6

This sample solution is based on choices of 5 cm for the radius of the base and 200 cm³ (or 200 mL) for the volume of the cone. Students may use numeric and/or algebraic reasoning to find the height of the pyramid:

$$h = \frac{3 \cdot V}{B} = \frac{3 \cdot V}{\pi \cdot r^2} = \frac{3 \cdot 200}{\pi \cdot 5^2} \cdot \frac{600}{\pi \cdot 25} = \frac{24}{\pi} \approx 7.6 \text{ cm}$$

By experimenting with paper or using mental visualization, students discover that the slant height of the cone is the radius of the circle that is needed to make the net for the lateral surface.

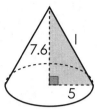

The diagram shows that you may calculate the slant height using the Pythagorean Theorem.

$$5^2 + \left(\frac{24}{\pi}\right)^2 = l^2$$

The resulting value for l is about 9.13 cm, which appears reasonable by comparison with the height of the cone. Students may now draw the circle for the lateral surface using a value slightly larger than 9.1 cm as the radius.

Next, they determine the size of the sector required to create the net for the lateral surface. Again using mental visualization or experimenting with paper, they discover that the circumference of the base must be equal to the arc length on the larger circle that they just drew. (This ensures that the base and the lateral surface will "match up" properly when joined.)

Finding the central angle that will produce the desired arc length is often a significant challenge for students. The solution involves using proportional reasoning based on the ratio of the circumferences of the circles. The base has a circumference of $2 \cdot \pi \cdot 5 = 10 \cdot \pi \approx 31.4$ cm. The larger circle has a circumference of approximately $2 \cdot \pi \cdot 9.13 = 18.26 \cdot \pi \approx 57.4$ cm. The ratio of the circumferences is about:

Teacher's Note. By using their knowledge of similar figures or by writing the ratio in a simpler form, $\dfrac{2 \cdot \pi \cdot 5}{2 \cdot \pi \cdot 9.13} = \dfrac{5}{9.13} \approx 0.548$, some students may discover that they can use the ratio of the radii instead of the circumferences!

$$\frac{31.42}{57.37} \approx 0.548 = 54.8\%$$

Therefore, the central angle that creates the necessary arc length is approximately 54.8% of 360°:

$$0.548 \times 360° = 197°$$

151

Based on these calculations, students' drawings will look like those below.

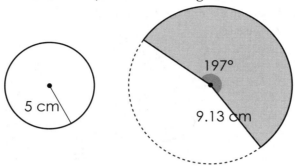

Students cut out the shaded sector of the large circle, fold it, and tape it to form the lateral face of the cone. (It is not necessary to cut out or attach the base.) They may test their work two ways:

1. Place the lateral surface over the drawing of the base to ensure that the two fit together properly.
2. Measure out 200 mL of sand or popcorn and pour it into the cone to test that it has the correct volume.

If students find that their cone does not pass the tests, it may be that they used the smaller sector of the circle (the part shown in white in the picture above). If this is not the problem, they will have to search for other sources of error and try to build it again.

To find the area of the lateral surface, students may simply calculate the area of the sector that they folded to form it! This is approximately 54.8% of the area of the larger circle:

$$S \approx 0.548 \times \left(\pi \times 9.13^2\right) \approx 143.5 \text{ cm}^2$$

Problem #7

7. Predict and test a formula for the lateral surface area of a cone. If your formula does not work, continue predicting and testing until you find one that does. Prove that your formula applies to all cones.

Questions and Conversations for #7

» *How can you distinguish an area formula from a volume formula, even if they are unfamiliar?* Compare and contrast as many familiar examples of area and volume formulas as you can. (Students should eventually discover that area formulas always contain a product of exactly two lengths, while volume formulas always have a product of three lengths.)

» *What else might an area or volume formula contain?* It may also involve multiplication by one or more numbers (including π).

» *What variables are likely to be in a formula for the lateral surface area of a cone?* Think about the length measurements that you can make for a cone.

(Possibilities include the radius of the base, the height or the slant height of the cone, or even the circumference of the base.)

» *Once you have an idea for a formula or expression, how can you test it?* Check to see that it gives you the correct answer for Problem #6. Create other cones, and test it on them as well.

» *Why doesn't repeated testing of the formula prove that it is correct? What can you do to be certain that the formula* always *works?* The formula may fail to work the next time you test it! In order to be certain, you must be able to show or explain *why* it works.

Solution for #7

The correct formula for the lateral surface area (S) of a cone is $S = \pi \cdot r \cdot l$.

An area formula for a figure always involves the product of exactly two lengths. For a cone, the most obvious lengths to try are the radius of the base (r), the height (h), or the slant height (l). There are many possible combinations of two of these lengths:

$$r \cdot h, \ r \cdot l, \ l \cdot h, \ r \cdot r, \ l \cdot l, \text{ or } h \cdot h$$

In order to capture information about the entire cone, you need at least one length that tells you about the size of the base and one that tells you something about how tall it is. This leaves only two possibilities, $r \cdot l$ or $r \cdot h$.

You should probably also expect the expression to contain the number π, because cones contain circles. If you insert a factor of π, you get $\pi \cdot r \cdot l$ or $\pi \cdot r \cdot h$. You should consider the possibility that the expression contains other numbers as well. However, if you test these expressions, you find that $\pi \cdot r \cdot l$ gives you the correct answer to Problem #6!

Of course, to be confident that this expression always works, you need to test it on more examples, or explain why it works. If you use variables in place of numbers, your thinking process in Problem #6 will guide you toward a method for proving it.

The area of the sector that you folded to form the lateral surface was a fraction of the area of the larger circle. This fraction was the ratio of the circumferences, which is the same as the ratio of the radii, $\dfrac{r}{l}$. Multiplying this fraction by the area of the large circle gives:

$$\frac{r}{l} \cdot \left(\pi \cdot l^2 \right)$$

Students who know some algebraic procedures may use them to simplify this expression. Other students may have to play with it for a while to see that it is equivalent to $\pi \cdot r \cdot l$. For example, they may realize that multiplying two factors of l and then dividing by l has the same effect as simply multiplying once by l (because dividing by l undoes multiplication by l).

153

WRAP UP

Share Strategies

Give students an opportunity to compare answers and methods. Ask them to discuss the precision in their answers. How many decimal places did they keep? Did other students obtain the same results for these decimal places? If not, what caused any discrepancies? What does this suggest about the number of decimal places that should be reported?

Summarize

Answer any remaining questions that students have. Summarize key ideas:
» The volumes of prisms and cylinders are given by the formula $V = B \cdot h$. This formula implies that: (1) When the area of the base changes by some factor, the volume changes by the same factor, and (2) if you "tilt" a prism (making it *oblique*) while maintaining its height, the volume does not change.
» The volumes of pyramids and cones are given by the formulas $V = \frac{1}{3} \cdot B \cdot h$.

The two statements above regarding prisms and cylinders continue to apply in this case. You may use nets to help you understand what causes the factor of $\frac{1}{3}$.
» Lateral surface areas of prisms, cylinders, pyramids, and cones generally change when you "tilt" them. Lateral surface areas of *oblique* figures are often substantially more difficult to calculate.
» Nets can be helpful for visualizing and analyzing three-dimensional figures. They are especially useful for understanding and calculating surface area.
» When carrying out a series of calculations, it is best to keep intermediate results in exact form for as long as it is practical to do so—delaying the use of a calculator until later in the process. When it is necessary to round intermediate results, keep more decimal places than you expect to report in your final result. If you are unsure how the precision of your result has been affected by rounding during the calculation process, you can learn a great deal by repeating the calculations while making different rounding choices along the way.

Further Exploration

Ask students to think of new questions to ask or ways to extend this exploration. Here are some possibilities:
» Extend Problem #4 by using bases other than squares to create pyramids whose apex is directly above a vertex of the base. (Try equilateral triangles,

pentagons, etc.) Are you able to use your constructions to find the volumes or surface areas? If so, do it. If not, explain what prevents you from doing this.

» Verify your formula for the lateral surface area of a cone by starting with pyramids whose base is a regular polygon. What happens to the surface area when this polygon has more and more sides?

» Create new formulas for the lateral surface area of a cone using a different combination of variables. For example, try to create a formula that includes h (the height of the cone) or c (the circumference of the base).

Exploration 9
Measuring Oceans

INTRODUCTION

Materials

» Scientific calculator

Prior Knowledge

» Read and write algebraic expressions that omit multiplication symbols.
» Know and use the standard formulas for the volume and surface area of a sphere.
» Understand and use scientific notation to read, write, and calculate with large numbers (recommended).
» Understand the Pythagorean Theorem and the distributive property. (Problem #6)

Learning Goals

» Attend to the precision of numbers when doing calculations and reporting results.
» Apply formulas for the surface area and volume of a sphere to solve challenging problems.
» Devise and apply strategies such as *guess-test-revise* or *work backward* to solve equations (see Problem #3).
» Generalize the concept of an *average* to cases involving continuously changing measurements (see Problem #4).
» Use spatial visualization skills, algebraic thinking, and logical reasoning to derive a formula for the volume of a sphere (see Problem #5).
» Communicate complex mathematical ideas clearly.
» Persist in solving challenging problems.

Launching the Exploration

Motivation and purpose. To students: Sometimes, you may be surprised by the far-reaching conclusions you can draw from relatively little information. In this activity, you apply your knowledge of formulas for the surface area and volume of a sphere to estimate the capacity of Earth's oceans. However, your task goes beyond mere calculation. You need to think about what information is required and how precisely you know the answer.

Understanding the problem. Look through the exploration to give students a sense of what it involves. In Stage 1, they answer a question about the Earth's oceans in two ways, using a volume formula and using a surface area formula. The question is very open-ended. Part of the students' task is to think about what information they might gather to use in their calculations.

Parts of Stage 2 may actually be less challenging than Stage 1, but students extend the ideas from their work in the first two questions to think more deeply about the formulas and their results. Stage 3 is very challenging for most students. Here, they think their way through a method for deriving the formula for the volume of a sphere.

To prepare for Stage 1, begin with a brief discussion of the meaning of an *order of magnitude estimate*. (See the Questions and Conversations for #1.) Just for fun, ask students to make a guess (using a place value name or a power of 10) so that they have something to compare their answer to when they have completed their work.

Next, solicit ideas for data that could be gathered in order to answer the question. If students come up with suggestions different from those used in the Solutions, consider using their ideas at first—unless, of course, someone suggests simply looking the answer up! Then redo the problem using the data suggested in this book. Compare the results and the solution methods.

NAME: _____ DATE: _____

STUDENT HANDOUT

Volume of a Sphere: $V = \dfrac{4}{3}\pi r^3$ **Surface Area of a Sphere:** $S = 4\pi r^2$

Stage 1

1. Use one of the formulas above along with some additional data to find an *order of magnitude estimate* of the amount of water in the Earth's oceans. Show your work, and explain your thinking. Discuss factors that might affect the accuracy of your results.

2. Show a method for performing the calculation in Problem #1 using the formula that you did not use the first time. Compare the two methods. Is one of them more efficient or easier to understand? Do they produce the same result?

Stage 2

3. If it were possible to have a planet made entirely of water, how large would it have to be in order to contain all of the water in the Earth's oceans? Is this a realistic size for a planet?

4. Invent, describe, and justify a method to determine the average depth of a rectangular swimming pool that has a floor and walls with the cross section below. Each square on the grid represents one square foot. What does your method suggest about the meaning of the average depth of the Earth's oceans?

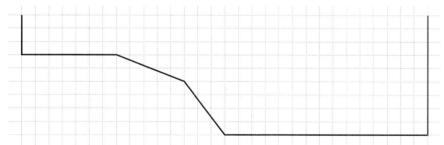

If two three-dimensional figures lie between two parallel planes, and every plane parallel to them intersects both figures in cross sections of equal area, then the two figures have equal volumes. This is known as *Cavalieri's Principle*, and it is a three-dimensional version of the concept from the Introduction to Exploration 8: Designing Nets. (See the discussion of triangle area on page 136.)

5. The diagrams below show a cylinder with a cone removed (left) and a hemisphere (right). The heights of the cylinder and cone are equal to the radius of the sphere. Apply Cavalieri's Principle to these diagrams in order to show why the volume of a sphere is determined by the formula $V = \frac{4}{3}\pi r^3$.

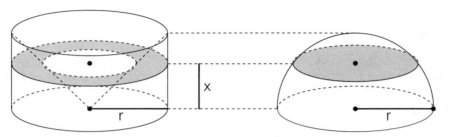

Advanced Common Core Math Explorations: Measurement & Polygons © Prufrock Press Inc.

TEACHER'S GUIDE

STAGE 1

Volume of a Sphere: $V = \frac{4}{3}\pi r^3$ **Surface Area of a Sphere:** $S = 4\pi r^2$

Problem #1

1. Use one of the formulas above along with some additional data to find an *order of magnitude estimate* of the amount of water in the Earth's oceans. Show your work, and explain your thinking. Discuss factors that might affect the accuracy of your results.

Questions and Conversations for #1

This section contains ideas for conversations, mainly in the form of questions that students may ask or that you may pose to them. Be sure to allow students to do most of the thinking and talking!

» *What is an* order of magnitude estimate? An *order of magnitude estimate* determines a value to the nearest power of ten (or equivalently, the nearest place value). See Exploration 4: Million, Billion, Trillion from the *Numbers and Operations* book in this series to explore this concept in more depth.

» *What data in addition to the diameter or radius of the Earth are needed to answer the question?* It would be sufficient to know the average depth of the Earth's oceans and the percent of the Earth's surface covered by water.

> **Teacher's Note.** If students suggest finding and combining data for each ocean individually, allow them to pursue this approach. They could later compare methods.

» *Could you estimate the percent of the Earth's surface covered by water without looking it up?* Although students may look this up if they like, they could learn a lot by thinking of ways to do it themselves. Some may suggest covering a globe or map with squares of equal area and calculating the percentage of squares that cover the water. If they use certain types of maps, they may find that areas near the North and South Poles appear to be much too large. In fact, there are many ways to map a sphere onto a flat surface (using different *projections*), but all of them distort some measurements. The commonly used *Mercator* projection exaggerates area measurements near the poles. Some students may be interested in searching for a map of the Earth that uses an *equal-area projection*.

» *How would you determine the average depth of an ocean (or of all the oceans)?* Although it will not be practical for students to do this themselves, they may have ideas about how it could be done. For example, they may suggest taking depth measurements at different locations and averaging the results. Ask them to think about what could be done to increase the likelihood of an accurate result. (Possibilities include taking a large number of measurements and choosing locations randomly.)

» *What units should you use to express the volume of the Earth's oceans?* Any unit of volume is acceptable. Some students may suggest using large units such as cubic miles or cubic kilometers due to the large size of the Earth.

Teacher's Note. Units of *liters* are used in the solution below, because they are a familiar metric unit of volume. The amount of water is so great that choosing a larger unit such as cubic meters or even cubic kilometers would still produce a very large number for an answer.

Solution for #1

An order of magnitude estimate for the volume of the Earth's oceans is 10^{21} liters (on the low end of this power of 10).

Sample strategy. This solution makes use of the volume formula for a sphere. It shows what happens if students use customary units for data in spite of the fact that they want to express their results in metric units. (Those who choose units that are more easily related to the units in which they intend to express their answer will have less work to do!) The values used here are 3960 miles for the Earth's radius, 2.5 miles for the average depth of the oceans, and 70% for the portion of the Earth's surface covered by water.

A typical strategy is to find the volume of a "shell" of water at the Earth's surface. By considering the Earth's radius with and without a 2.5 mi layer of water:

Teacher's Note. If students wait to make the conversions to metric units until after they have calculated the volumes, they will face the challenge of converting cubic units. For example, although there are 12 inches per foot, there are $12^3 = 1728$ cubic inches per cubic foot. Similarly, there are 2.54 cm per inch but $2.54^3 = 16.387064$ cubic centimeters per cubic inch. This is a common source of error.

» 1 mi = 5280 ft $\times$ 12 in per foot $\times$ 2.54 cm per in = 160,934.4 cm
» Radius of the Earth: 3960 mi $\times$ 160,934.4 cm per mile $\approx 6.3730 \times 10^8$ cm
» "Inner" radius: $(3960 - 2.5)$ mi $\times$ 160,934.4 cm per mi $\approx 6.3690 \times 10^8$ cm
» Volume of Earth: $V = \dfrac{4}{3}\pi \times \left(6.3730 \times 10^8\right)^3 \approx 1.0842 \times 10^{27}$ cm^3
» Inner volume: $\dfrac{4}{3}\pi \times \left(6.3690 \times 10^8\right)^3 \approx 1.0822 \times 10^{27}$ cm^3
» Volume of shell:
$$\left(1.0842 \times 10^{27}\right) - \left(1.0822 \times 10^{27}\right) = 0.0020 \times 10^{27} = 2.0 \times 10^{24} \text{ cm}^3$$

» Amount of water in the shell: $0.7 \times \left(2 \times 10^{24}\right) = 1.4 \times 10^{24}$ cm^3
» Value in liters: 1.4×10^{24} cm$^3 \div 1000$ cm^3 per liter $= 1.4 \times 10^{21}$ liters.

Students may wonder how much of a difference it makes that the original data was not extremely precise. Or they could question the effect of any rounding that they did. Encourage them to experiment by using slightly different initial data, or keeping more or fewer decimal places as they work. How much do these changes affect the results?

Because the radius of the Earth is not exactly 3960 miles (in fact, the Earth is slightly wider at the equator than at the poles!), students may wonder how much it matters whether you add or subtract the 2.5 mile ocean depth to or from the Earth's radius. Again, try it both ways. They should find that it makes little difference.

Students may also question whether they should have considered the effects of other bodies of water. Noticing the small surface area of lakes, seas, rivers, etc. relative to that of the oceans combined with their much smaller depth would suggest that it is not a significant consideration when making an order of magnitude estimate.

Problem #2

2. Show a method for performing the calculation in Problem #1 using the formula that you did not use the first time. Compare the two methods. Is one of them more efficient or easier to understand? Do they produce the same result?

Questions and Conversations for #2

The Questions and Conversations for Problem #1 apply to this problem as well.

» *How might you use a surface area measurement to estimate a volume?* Think about the process of finding the volume of a prism. Is it possible to apply a similar idea here? Why or why not?

Solution for #2

An order of magnitude estimate of the volume of the Earth's oceans is 10^{21} liters (on the low end of this power of 10).

Sample strategy. This solution makes use of the formula for the surface area of a sphere. The process may be more comfortable for some students than the method shown in Problem #1, because it involves multiplying an area (the surface of the Earth) by a height (the depth of the oceans). The procedure is similar to that of finding the volume of a prism, but does it make sense to apply it to this situation? Imagine cutting and "flattening" an imaginary shell of water (see the Solution for Problem #1) on the surface of the Earth to form a prism so that the inner shell forms one base, and the outer shell forms the other. Because two bases are not exactly the same size, some distortion will occur in the process of flattening. However, because the depth of the shell is small compared to the radius of the Earth, the relative sizes of the bases are quite close, and it is possible to form something that looks something like a prism! (Compare the volume of the shell below to the corresponding calculation in Problem #1.)

> » 1 mi = 5280 ft × 12 in per foot × 2.54 cm per in = 160,934.4 cm
> » Radius of the Earth: 3960 mi × 160,934.4 cm per mile ≈ 6.3730×10^8 cm
> » Average depth of the oceans: 2.5 mi × 160,934.4 cm per mile = 402,336 cm
> » Surface area of the Earth: $S = 4\pi r^2 = 4\pi \times \left(6.3730 \times 10^8\right)^2 \approx 5.104 \times 10^{18}$ cm^2
> » Volume of shell: $\left(5.104 \times 10^{18}\right) \times 402,336 = 2.054 \times 10^{24}$ cm^3
> » Amount of water in the shell: $0.7 \times \left(2.054 \times 10^{24}\right) \approx 1.4 \times 10^{24}$ cm^3
> » Value in liters: 1.4×10^{24} cm^3 ÷ 1000 cm^3 per liter = 1.4×10^{21} liters.

The answers using the methods in Problems #1 and #2 agree to two *significant digits* (the two digits in the first part of the scientific notation expressions).

STAGE 2

Problem #3

3. If it were possible to have a planet made entirely of water, how large would it have to be in order to contain all of the water in the Earth's oceans? Is this a realistic size for a planet?

Questions and Conversations for #3

> » *Can you answer the question simply by stating that the volume of the planet is approximately* 1.4×10^{21} *liters?* Yes. Of course, that does not make a very interesting problem! Furthermore, there are measurements other than volume that would probably do a better job of helping you visualize the size of the planet. What kinds of measurements might be more helpful?
> » *How can you begin if you don't know a series of "steps" to take?* Try making a reasonable guess, testing it, and using the result to help you improve your guess.

Solution for #3

The object would have a radius of approximately 694 km or about 431 miles. This is close to the size of a dwarf planet and is about 40% of the radius of Earth's moon.

In order to obtain an answer in units of centimeters, it may be easier to begin with:

$$1.4 \times 10^{21} \text{ liters} = 1.4 \times 10^{24} \text{ cm}^3$$

The question involves solving the equation $\frac{4}{3}\pi r^3 = 1.4 \times 10^{24}$. Some students will use a guess-test-revise strategy—choosing values for r, seeing if they work, and when they don't, using the results to inform the next guess.

Other students will "think backward."

(1) The value of πr^3 is $(1.4 \times 10^{24}) \div \frac{4}{3} \approx 1.05 \times 10^{24}$ cm³. (Students may instead multiply by $\frac{3}{4}$ or multiply by 3 and divide by 4.)

(2) The value of r^3 is approximately $(1.05 \times 10^{24}) \div \pi \approx 3.34 \times 10^{23}$ cm³.

(3) Some students may use a guess-test-revise process to determine that $r \approx 6.94 \times 10^7$ cm, which is equal to 694 km or about 431 miles. If they are familiar with the concept of a cube root (and know how to use a calculator to find it), they may calculate $\sqrt[3]{3.34 \times 10^{23}} \approx 6.9 \times 10^7$ directly.

Problem #4

4. Invent, describe, and justify a method to determine the average depth of a rectangular swimming pool that has a floor and walls with the cross section below. Each square on the grid represents one square foot. What does your method suggest about the meaning of the average depth of the Earth's oceans?

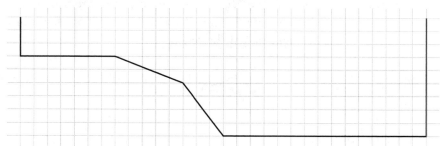

Questions and Conversations for #4

» *Does it matter that the deepest part of the pool is longer than the other parts?* Yes. Longer sections of the pool hold more water for a given depth, so they should have a greater effect on the average depth calculation.

» *Does it make sense to find the average of the average depths for each of each of the four sections of the pool? Why or why not?* No, because this fails to take into account the different sizes of the sections. (Students who make this error will perform the calculation $(3 + 4 + 7 + 9) \div 4$, resulting in an incorrect answer of 5.75 feet.)

» *How can you account for the sizes of the different sections in your calculations?* There are at least a couple of options. (1) Consider adjusting the average depth of each section individually. (2) Think of ways to use volumes.

» *Does it make sense to focus on areas instead of volumes? Why or why not?* Yes, because the cross section of the pool looks the same everywhere. The cross-sectional area times the width of the pool is equal to the volume of the pool.

Solution for #4

The average depth is $6\frac{17}{30} \approx 6.6$ ft.

Sample strategy 1. Use a *weighted* average. Give the average depth of each section the proper "weight" by multiplying it by its fraction of the length of the entire pool. The average depths (from left to right) are 3 feet, 4 feet, 7 feet, and 9 feet. The 3-foot section is $\frac{7}{30}$ of the length of the pool, etc.:

$$\left(3 \cdot \frac{7}{30}\right) + \left(4 \cdot \frac{5}{30}\right) + \left(7 \cdot \frac{3}{30}\right) + \left(9 \cdot \frac{15}{30}\right) = \frac{21}{30} + \frac{20}{30} + \frac{21}{30} + \frac{135}{30} = \frac{197}{30} = 6\frac{17}{30} \text{ ft}$$

Sample strategy 2. Find the total area of the diagram and divide the result by the length of the pool. The total area is 197 ft². $197 \div 30 \approx 6.6$ ft. This strategy captures an important meaning of the average depth. If another pool of the same length and width had this constant depth, it would hold the same amount of water as the original pool.

Imagine that the floors of all the oceans were smoothed out to the same depth so that the oceans still held the same volume of water. The constant depth of these imaginary oceans is the *average depth* of the real oceans!

STAGE 3

Problem #5

5. The diagrams below show a cylinder with a cone removed (left) and a hemisphere (right). The heights of the cylinder and cone are equal to the radius of the sphere. Apply Cavalieri's Principle to these diagrams in order to show why the volume of a sphere is determined by the formula $V = \frac{4}{3}\pi r^3$.

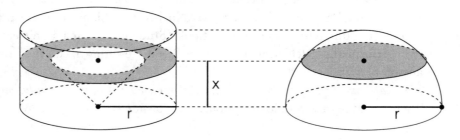

Questions and Conversations for #5

» *What happens to the areas of the cross sections (the circles) as you move from the bottom to the top of the hemisphere?* They become smaller, because the radii of the circles get smaller.

» *What happens to the areas of the cross sections (the rings) as you move from the bottom to the top of the cylinder with the cone removed?* They become smaller because the outer edges of the rings remain the same while the inner edges move outward from the center, leaving larger circular holes in the middle.

» *How might the area of a circle and a ring at the same height compare?* It appears that they could be equal, because they are the same at the bottom of each figure (πr^2) and at the top (0), and they both become smaller as they move up.

» *How can you verify that the cross sections have (or do not have) the same area at each height?* Use algebraic techniques to calculate formulas for the cross sections at each height.

» *What might you add to the drawings in order to determine the areas of the cross sections?* Consider beginning by marking important measurements that you will need such as the radii of some circles. What additional features might you add to help you calculate their lengths?

Solution for #5

Begin by proving that the cross sections at each height have the same area.
A strategy to find the cross-sectional area for the hemisphere:
Begin by drawing some auxiliary segments to form a right triangle.

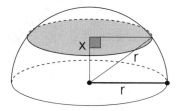

Use the Pythagorean Theorem with this triangle to find that the square of the radius of the cross-sectional circle is $r^2 - x^2$. Therefore, the area of this circle is $\pi\left(r^2 - x^2\right)$.

A strategy to find the cross-sectional area for cylinder with the cone removed:
As before, begin by drawing auxiliary segments to form a right triangle.

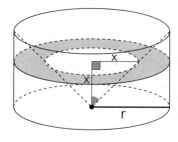

The two legs of the right triangle have the same length (x), because the radius of the base of the cylinder is equal to the height of the cylinder, making the acute angle marked at the bottom of the diagram have a measure of 45°. Because the radius of the outer circle of the ring is r, and the radius of the inner circle is x, the area of the ring is $\pi r^2 - \pi x^2$.

The distributive property shows that the areas of the two cross sections are equal:

$$\pi\left(r^2 - x^2\right) = \pi r^2 - \pi x^2$$

According to Cavalieri's Principle, the volumes of the two figures are equal. Therefore, you can find the volume of the hemisphere by subtracting the volume of the cone from the volume of the cylinder in the left figure:

volume of cylinder $-$ volume of cone $=$

$$Bh \quad - \quad \frac{1}{3}Bh \ =$$

$$\left(\pi r^2\right)\cdot r \quad - \quad \frac{1}{3}\left(\pi r^2\right)\cdot r \ =$$

$$\pi r^3 \quad - \quad \frac{1}{3}\pi r^3 \ =$$

$$\frac{2}{3}\pi r^3$$

To obtain the volume of the sphere, simply double the volume of the hemisphere!

$$V = 2\cdot\frac{2}{3}\pi r^3 = \frac{4}{3}\pi r^3$$

WRAP UP

Share Strategies

Give students an opportunity to compare their answers to Problems #1 and #2. If there are disagreements, determine whether there was a conceptual or calculation error involved. Then look at the level of precision that students reported. To what place value do all of the answers agree? What are the sources of any discrepancies? What does all of this imply about the appropriate number of decimal places to report in your answer?

Ask students to share and compare strategies for any additional problems that they completed. Discuss advantages and disadvantages of each.

Summarize

Answer any remaining questions that students have. You may also want to summarize and expand on a few key ideas:

» Combining visual reasoning with knowledge of formulas when solving measurement problems may provide new insights and suggest helpful strategies.

» When reporting results of calculations, you should not show more decimal places than you know with a reasonable level of confidence. In making this determination, you need to consider the precision of the initial measurements and any rounding that you may have done in the calculation process. When it is difficult to determine how precisely you know the result, consider analyzing the effects of making small changes in the values of the initial measurements or doing the calculations using a variety of strategies.

» The strategy shown in the solution to Problem #5 for deriving the volume formula for a sphere is based on an idea originally described by the Greek mathematician Archimedes.

Further Exploration

Ask students to think of new questions to ask or ways to extend this exploration. Here are some possibilities:

» Use the volume formula for a sphere to derive a formula for its surface area. (Imagine filling the sphere with identical pyramids whose apexes lie at the center of the sphere. Imagine having more and more of these pyramids as you make them smaller and smaller. What can you say about the sum of their bases?)

» Do some research to learn more about map projections. How are geometry and algebra used to create them? Why are some measurements always distorted? What are some of the advantages and disadvantages of various projections? How can you choose a projection that is well suited to a given situation?

» In Problems #1 and #2, you had to think about the data you would need to gather in order to answer a question. Take this idea a step further by creating and solving open-ended problems for situations in which the values of the data are unknown or highly variable. Sometimes, you may be able to use popular expressions for inspiration! For example, how much worse does a situation become when you *make a mountain out of a molehill?* How many tears does it take to *cry a river?* How long would it take to *move a mountain?*

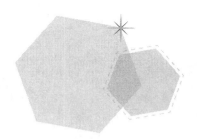

References

Bell, M., Bretzlauf, J., Dillard, A., Hartfield, R., Isaccs, A., McBride, J., . . . Saecker, P. (2007). *Everyday mathematics: Teacher's lesson guide, grade 6 volume 1* (3rd ed.). Chicago, IL: McGraw Hill.

Frayer, D. A., Frederick, W. C., & Klausmeier, H. G. (1969). *A schema for testing the level of concept mastery* (Working paper No. 16.). Madison: University of Wisconsin.

Fuys, T., Geddes, D., & Tischler, R. (1988). The Van Hiele model of thinking in geometry among adolescents [Monograph]. *Journal for Research in Mathematics Education, 3,* 34.

International GeoGebra Institute. (2014). *GeoGebra.* Retrieved from http://www.geogebra.org

National Governors Association Center for Best Practices, & Council of Chief State School Officers. (2010). *Common core state standards for mathematics.* Washington, DC: Authors.

Sheffield, L. J. (2003). *Extending the challenge in mathematics: Developing mathematical promise in K–8 students.* Thousand Oaks, CA: Corwin Press.

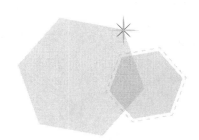

About the Author

Jerry Burkhart has been teaching and learning math with gifted students in Minnesota for nearly 20 years. He has degrees in physics, mathematics, and math education from University of Colorado, Boulder, and Minnesota State University, Mankato. Jerry provides professional development for teachers and is a regular presenter at conferences addressing topics of meeting the needs of gifted students in mathematics.

Common Core State Standards Alignment

Exploration	Common Core State Standards in Mathematics
Exploration 1: Polygon Perambulations	4.MD.C Geometric measurement: understand concepts of angle and measure angles. 7.G.B Solve real-life and mathematical problems involving angle measure, area, surface area, and volume. 8.G.A Understand congruence and similarity using physical models, transparencies, or geometry software.
Exploration 2: Impossible Polygons	4.G.A Draw and identify lines and angles, and classify shapes by properties of their lines and angles. 5.G.B Classify two-dimensional figures into categories based on their properties. 7.G.A Draw construct, and describe geometrical figures and describe the relationships between them.
Exploration 3: Starstruck!	7.G.B Solve real-life and mathematical problems involving angle measure, area, surface area, and volume. 7.EE.A Use properties of operations to generate equivalent expressions. 5.OA.A Write and interpret numerical expressions. 5.OA.B Analyze patterns and relationships. 5.G.A Graph points on the coordinate plane to solve real-world and mathematical problems.
Exploration 4: Geoboard Squares	6.G.A Solve real-world and mathematical problems involving area, surface area, and volume. 7.G.B Solve real-life and mathematical problems involving angle measure, area, surface area, and volume. 6.EE.A Apply and extend previous understandings of arithmetic to algebraic expressions.
Exploration 5: Creating Area Formulas	6.G.A Solve real-world and mathematical problems involving area, surface area, and volume. 7.G.B Solve real-life and mathematical problems involving angle measure, area, surface area, and volume.

Exploration	Common Core State Standards in Mathematics
Exploration 6: A New Slant on Measurement	6.G.A Solve real-world and mathematical problems involving area, surface area, and volume. 7.G.B Solve real-life and mathematical problems involving angle measure, area, surface area, and volume. 8.G.B Understand and apply the Pythagorean Theorem. 7.EE.A Use properties of operations to generate equivalent expressions.
Exploration 7: Ladders and Saws	5.G.B Classify two-dimensional figures into categories based on their properties. 7.G.B Solve real-life and mathematical problems involving angle measure, area, surface area, and volume. 8.G.A Understand congruence and similarity using physical models, transparencies, or geometry software. 6.G.A Solve real-world and mathematical problems involving area, surface area, and volume.
Exploration 8: Designing Nets	6.G.A Solve real-world and mathematical problems involving area, surface area, and volume. 7.G.B Solve real-life and mathematical problems involving angle measure, area, surface area, and volume. 8.G.C Solve real-world and mathematical problems involving volume of cylinders, cones, and spheres.
Exploration 9: Measuring Oceans	8.G.C Solve real-world and mathematical problems involving volume of cylinders, cones, and spheres. 8.EE.A Work with radicals and integer exponents. 7.G.A Draw construct, and describe geometrical figures and describe the relationships between them. 7.G.B Solve real-life and mathematical problems involving angle measure, area, surface area, and volume. 7.EE.B Solve real-life and mathematical problems using numerical and algebraic expressions and equations. 8.G.B Understand and apply the Pythagorean Theorem.

Note. Please see p. 7 of the book for details on how to connect and extend the core learning of content in these lessons.